Effective Math Instruction

Shifting to Meet Today's Standards

Author
Jared DuPree, Ed.D.

Foreword
Ruth Harbin Miles, Ed.S.

SHELL EDUCATION

Publishing Credits

Corinne Burton, M.A.Ed., *President*; Kimberly Stockton, M.A.Ed., *Vice President of Education*; Conni Medina, M.A.Ed., *Managing Editor*; Sara Johnson, M.S.Ed., *Content Director*; Paula Makridis, M.A.Ed. *Editor*; Kyleena Harper, *Assistant Editor*; Lee Aucoin, *Multimedia Designer*; Kevin Pham, *Production Artist*

Image Credits

p.168, 258 iStockphoto; all other images Shutterstock

Shell Education
5301 Oceanus Drive
Huntington Beach, CA 92649-1030
http://www.tcmpub.com/teachers
ISBN **978-1-4258-1508-0**

Table of Contents

Foreword

In 2009, the National Governor's Association and the Council of Chief State School Officers met to discuss the development of common standards to be used in all states to address the quality of mathematics instruction and student learning. As a result of that meeting, the Common Core Mathematics Standards were written as a set of shared goals and expectations to address knowledge and skills needed by students at each grade level to prepare them to succeed in college, career, and life.

As school districts work to implement the Common Core Mathematics Standards and the Standards for Mathematical Practice, some confusion and questions still exist. Some school districts are wondering whether their curriculum is rigorous enough to meet the Standards, how they can build coherence across the Standards, and how to identify and fill in gaps in their curriculum. In addition, some teachers and administrators want to know what types of questions should be asked to prepare students for assessments.

Effective Math: Shifting to Meet Today's Standards successfully addresses rigor and coherence, and how to identify curriculum gaps and prepare students for assessments. The book focuses on how teachers can increase their knowledge of the Common Core State Standards, which obviously is the key to students increasing their understanding of mathematics. The author has effectively laid out a series of lessons, templates, and exemplars for practical classroom application that make this book a great contribution to the field of mathematics education.

There are several unique features of the book that will be welcomed by teachers, mathematics coaches, and administrators. Among these are:

- a structural overview of the Common Core Content Standards and how to teach them, along with alignment between the Content Standards and the 8 Mathematics Practice Standards found in Chapter 1
- the instructional shift of coherence found in Chapter 3 that shows mathematics as a progression with interconnected topics students must clearly understand versus merely memorize
- the instructional shift of rigor in Chapter 4 that includes examples of rigorous problems along with tools for monitoring instructional rigor

I believe readers will find this publication to be an exemplary resource to support teacher understanding and instructional implementation of the Common Core Mathematics Standards, which will ultimately impact student achievement and prepare students for their futures.

—Ruth Harbin Miles
NCTM Board of Directors
Mathematics Educator
Mary Baldwin College, Staunton, Virginia

Acknowledgments

I would like to thank my family for their sacrifice during the writing of this book. Without your support, this book would not exist.

Thank you to my colleagues who provided critical feedback to help create a book that balances theoretical concepts with practical everyday classroom applications. You continue to push me, and I thank you for that.

Dr. Kaplan and Dr. Rousseau, you have left an indelible impression upon my professional beliefs. This book represents many ideas that I embrace as a result of your guidance. Knowledge is contested and constructed. Life is not always linear; it extends beyond the confines of a rigid box.

I would like to acknowledge all of the dedicated teachers and administrators who embrace the challenge of learning mathematics in order to teach mathematics in a meaningful way. May this book provide insight as you make meaning of mathematics and persevere.

Introduction

Understanding the Standards

The Common Core State Standards in Mathematics (CCSS-M) are learning goals that outline what students should know and be able to do in the discipline of mathematics. The standards were created to ensure that all students are college and career ready when they graduate from high school, an effort coordinated by the Council of Chief State School Officers (CCSSO) and the National Governors' Association (NGA) in collaboration with teachers, administrators, and mathematics experts. The expectation is that teachers will now incorporate Common Core measures into their instructional practice, which will ultimately prepare students for 21st century demands. Students must be able to think critically, work collaboratively, communicate effectively, think creatively, and make use of technological tools. How do teachers construct environments for students to acquire 21st century skills? This book is designed to help teachers and administrators answer this question. This book includes resources that enable teachers to increase their knowledge of the Common Core State Standards, which is key to students increasing their understanding of mathematics. Not only will teachers have the opportunity to increase their understanding of math content, but they will also learn how to develop standards-based lessons and measure students' understanding of the content using authentic assessment practices. The combination of all of these skills is often known as pedagogical content knowledge. Specialized content pedagogical knowledge is seen as a requisite for effective mathematics instruction (Van Zoest, Stockero, and Taylor 2011). This book is intended as a guide for effective Common Core mathematics instruction.

Parallel Sets of Standards

The CCSS-M represent the intersection of two parallel sets of standards. The content standards define the skills and knowledge in the discipline of mathematics. These standards were constructed on a series of progressions defined by the natural structure of mathematical concepts. Knowledge of students' developmental readiness was also taken into account. The content standards are delineated by grade level for grades kindergarten through eight. Conceptual categories are used at the high school level. The other set of standards that run parallel to the content standards are the standards for mathematical practice. The practice standards are actions, dispositions, and characteristics demonstrated by mathematicians as they negotiate the mathematics content. The practice standards are not a superordinate or subordinate set of standards. The two sets of standards are designed to work together. They should be taught concurrently.

Instructional Shifts

The CCSS-M define a set of mathematical expectations, but they do not dictate a prescribed curriculum. Comprehensive curricula include academic standards, targeted learning objectives, purposeful learning experiences, and the assessment of these experiences. In order to preserve the integrity of the standards during instructional delivery, alignment between the standards and curriculum should incorporate the three instructional shifts of focus, coherence, and rigor.

Focus describes the content that should be emphasized at each grade level. Certain content, commonly called "high-leverage content," has lasting implications for student readiness in mathematics. A majority of instructional time should be allocated to content denoted as "major clusters." The remaining time should then be used to teach supporting and additional clusters. As a result, students will have time to develop an understanding of integral mathematics topics at their unique grade level.

Coherence refers to the relationships among mathematical concepts. It highlights the importance of making connections between and across grade-level standards. All concepts presented should be situated in the larger context of math in order to construct schema and increase retention of information. Creating opportunities for students to see mathematics as a coherent subject eliminates the need for rote memorization of content.

Rigor describes the union of conceptual knowledge, procedural skills and fluency, and the application of knowledge in a real-world context. All three subsets should be represented with equal intensity to create a rigorous exploration of a mathematical concept. In order for students to understand mathematics and meet Common Core expectations, they must have an opportunity to be exposed to many different question types that target each component of rigor. Rigor is not reserved for students at a higher performance level.

The Common Core State Standards in Mathematics, along with the associated instructional shifts of focus, coherence, and rigor, are designed to develop mathematically proficient students. An understanding of what it means to be mathematically proficient may challenge the existing paradigm held by educators.

Assessment

The federal government established a common goal of proficiency in mathematics for all students through the policy of No Child Left Behind (US Department of Education 2001). This measure of mathematical proficiency has changed with the creation of the Common Core State Standards (CCSS). A new law, Every Student Succeeds Act (ESSA), signed by President Obama on December 10, 2015, focuses on preparing all students for success in college and careers.

In the past it has been noted by researchers such as Barbeau (1989) that district leaders often viewed mathematics as an established body of rules and procedures, and viewed "doing mathematics" as the mere manipulation of numbers. Previous state assessments reflected this underlying belief. The current assessments created by the Partnership for Assessment of Readiness for College and Careers (PARCC) and the Smarter Balanced Assessment Consortium (SBAC) reflect a more comprehensive view of "doing mathematics." The Common Core has established a new standard of proficiency. This standard includes a student's command of knowledge across multiple knowledge dimensions, in addition to procedural and factual knowledge, which was once the primary focus of mathematics education. Students must now possess conceptual knowledge and metacognitive knowledge.

Under the CCSS, mathematical proficiency is dependent upon the development of all four knowledge dimensions.

The illustration below shows the four knowledge dimensions. Different problem types generally target a specific knowledge dimension, but there is often overlap. Understanding discrete facts may require grouping into overarching ideas known as concepts. Application of concepts requires procedural skills. Reflection upon application and use of concepts incorporates metacognitive skills.

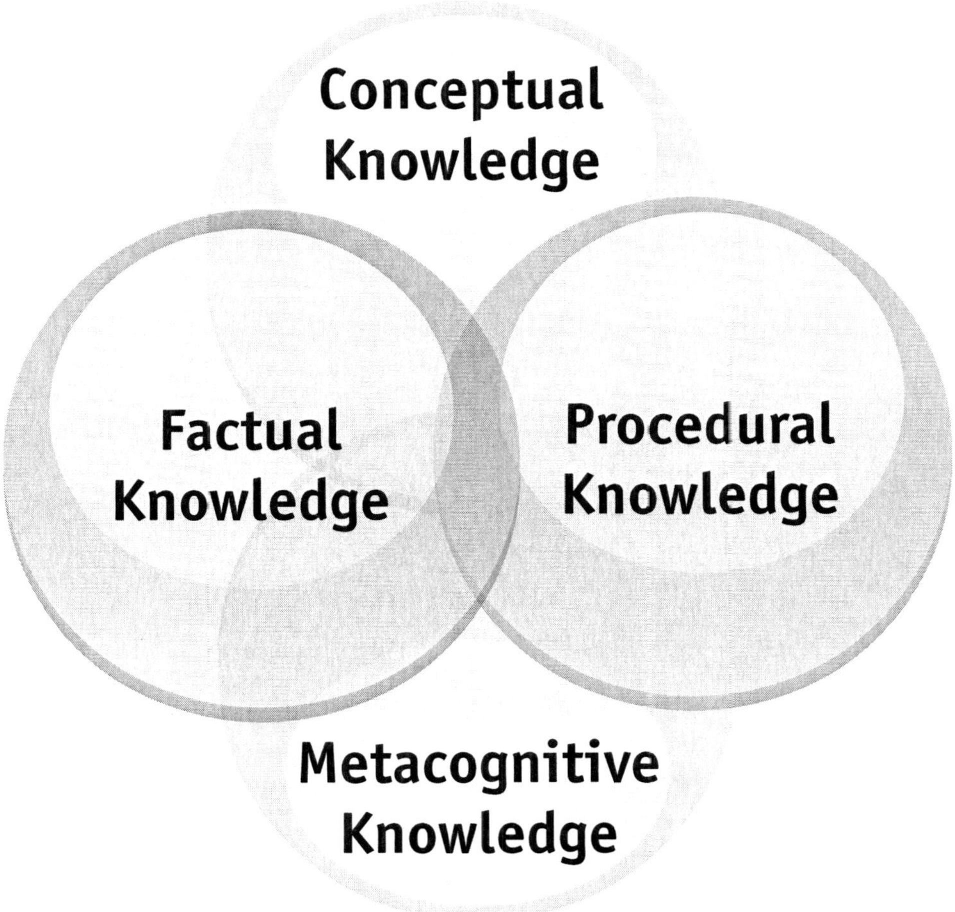

Each dimension of knowledge is defined and described in the chart on page 12. Different question types that target the development of each dimension are also included. The Common Core assessments created by PARCC and SBAC include multiple question types, such as selected response questions (SR) where students answer multiple choice questions, constructed response questions (CR) where students develop their own answer, extended response questions (ER) where students develop their answer and explain, technology enhanced questions (TE) where students use technological tools to answer questions, and performance tasks (PT) where students demonstrate conceptual and procedural knowledge in a real-world setting. A combination of these question types assesses a student's comprehensive understanding of mathematics knowledge across the four dimensions.

Dimension	Description	Question Example
Factual Knowledge	The basic elements that students must know to be acquainted with a discipline or solve problems in it • Knowledge of terminology • Knowledge of specific details and elements	What does "sum" mean? What is the value of the 8 in the number 83?
Conceptual Knowledge	The interrelationships among the basic elements within a larger structure that enable them to function together • Knowledge of classifications and categories • Knowledge of principles and generalizations • Knowledge of theories, models, and structures	What is the relationship between multiplication and division? How does the placement of a digit determine its value?
Procedural Knowledge	How to do something; methods of inquiry and criteria for using skills, algorithms, techniques, and methods • Knowledge of subject-specific skills and algorithms • Knowledge of subject-specific techniques and methods	What is the solution to this problem? $\frac{1}{2} + \frac{3}{4} = \square$ Find the product. $2 \times 3 \times 5 = \square$
Metacognitive Knowledge	Knowledge of cognition in general, as well as awareness and knowledge of one's own cognition • Strategic knowledge • Knowledge about cognitive tasks, including appropriate contextual and conditional knowledge • Self-regulating knowledge	Consider the following problem: Three friends have 18 jelly beans all together. They want to share the jelly beans equally. How many jelly beans should each friend get? What are multiple strategies that can be used to determine the solution? How is this problem familiar to something you've encountered? How did you determine what to do?

How to Use This Book

The purpose of this book is to provide practical tools for teachers and administrators to use in creating environments that lead to mathematically proficient students—students who will be able to demonstrate a command of all four knowledge dimensions.

Mathematical rigor in the classroom *for students* includes lessons that target conceptual knowledge, procedural knowledge, and the application of this knowledge in context. Professional development *for teachers* must also include opportunities for teachers to develop all three dimensions of rigor as it applies to the Common Core. This book primarily focuses on teachers' procedural knowledge of standards implementation as they apply the information and resources presented in this book. Procedural questions are targeted, such as:

- *How do I teach the Common Core State Standards in mathematics?*
- *How do I apply the information in this book to my own classroom practice?*

This book provides a series of lessons, templates, and exemplars for practical classroom application. It is essential to note that a teacher's conceptual understanding of the standards is necessary for educators to extend and innovate beyond the ideas in this book. Opportunities should also be provided for educators to construct their understanding of the Common Core Standards and lesson design through their own inquiry and investigation.

Teachers must understand the content standards and the mathematical practice standards in order to develop meaningful mathematics lessons. The units of study created should include the instructional shifts of focus, coherence, and rigor. These shifts will be explained throughout the book. Once lessons are designed that incorporate the Common Core elements, students must demonstrate their understanding through authentic assessment practices. The following illustrative framework represents the relationship between the standards, the Common Core shifts, lesson design, and assessment. The structure of this book and the connection between the chapters is designed based on this relationship.

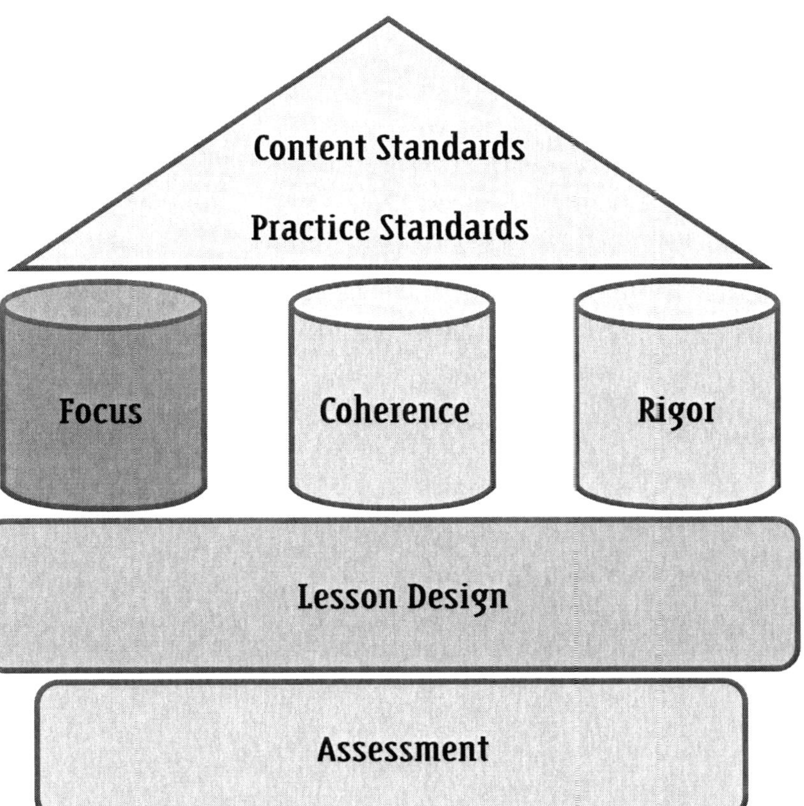

Overview of This Book

The following is an overview of the chapters in this book. A brief description captures the major ideas and resources in each chapter.

Chapter 1, Understanding the Standards, gives a structural overview of the content standards. In order to teach the standards, teachers must know the standards. Tools are provided to deconstruct and analyze the standards to develop a teacher's authentic understanding of the mathematic content.

This chapter also includes planning tools for incorporating the mathematical practice standards into curriculum design and instructional delivery. The alignment between the content standards and the math practice standards is explained. Questioning strategies that target the mathematics practice standards and illustrative prompts are also included.

Exploratory Questions:

1. *How do I increase my understanding of the standards?*
2. *What is the relationship between domains, clusters, and individual standards?*
3. *How do I determine which math practices to target in a lesson?*
4. *How do I teach students to understand the math practices alongside the math content?*

Chapter 2, Instructional Shift of Focus, highlights the grade-level foci and grade-level expected fluencies. Major and supporting clusters are defined and explained.

Exploratory Questions:

1. *How do I determine what content should be emphasized at my grade level?*
2. *How do I determine how much time should be allocated to content during curricular planning?*

In **Chapter 3, Instructional Shift of Coherence**, it shows how mathematics is a discipline in which all topics are interconnected. Students must understand these connections in order to learn and to not rely upon memorization. Resources are provided to help students make these vital connections.

Exploratory Questions:

1. *How do I use the math clusters to help students understand connections between lessons?*
2. *How might I create opportunities for students to understand how the supporting and/or the additional clusters connect to the major clusters?*

Chapter 4, Instructional Shift of Rigor, gives practical examples of problems that incorporate the three dimensions of rigor. Ideas for striking a balance among the dimensions of rigor are provided. Tools for monitoring instructional rigor are included.

Exploratory Questions:

1. *What are the qualities of a rigorous math lesson?*
2. *How do I determine what math problems to use in order to develop a particular dimension of rigor?*

In **Chapter 5, Lesson Design**, a variety of instructional models are provided that show how teachers can design standards-aligned lessons. An auditing tool to help teachers determine the Common Core readiness of their existing lessons is included.

Exploratory Questions:

1. *What are the features of a Common Core lesson?*
2. *How do I develop a unit that balances group investigation with direct instruction?*

Chapter 6, Assessment, explains the significant shift in how students must demonstrate their understanding of mathematics. This chapter provides rich examples of assessment question types.

Exploratory Questions:

1. *How will students be assessed using standardized testing measures?*
2. *How might I incorporate various question types in my assessment measures to provide opportunities for students to demonstrate their understanding of math in a meaningful way?*

This book is designed to help you develop a greater understanding of the Common Core State Standards and presents ways to incorporate this knowledge into your classroom practice. As you read this book, consider the relatedness between the different chapters. The questions presented above, in addition to many more, will be explored.

Chapter 1

Understanding the Standards

There are significant differences between the Common Core State Standards in Mathematics and previous state mathematics standards. One major difference is the set of parallel complementary standards that exist. The content standards are aligned with a second set of standards, known as the standards for mathematical practice. These two sets of standards are combined to present a comprehensive description of the CCSS-M. They not only describe what a student should know in mathematics, but also actions and ways of thinking that students should demonstrate while negotiating the mathematics content. This chapter highlights the relationship and interdependence between these two sets of standards. Both sets are a priority in developing students' meaningful understanding of mathematics.

This chapter includes the following tools that help teachers not only understand how to incorporate the standards into lesson design (procedural knowledge), but why (conceptual knowledge) and where (application in context) to implement this knowledge.

- **Content Standard Analysis Protocol**—This protocol helps teachers understand the content standards, highlighting terms and concepts that are unfamiliar.
- **Cluster Analysis Protocol**—Standards exist within overarching clusters. This tool develops a teacher's understanding of the relatedness between individual standards and the clusters that serve to create continuity within instruction.
- **Standards for Mathematical Practice Proficiency Guide**—This descriptive chart outlines the standards for mathematical practice, including a description of student actions that may indicate proficiency.
- **Standards for Mathematical Practice Analysis Protocol**—A comprehensive analysis of the mathematical practices should include an examination of what each standard means, how the standards relate to one another, and how they align with the content standards. This protocol provides an opportunity to understand standard alignment.
- **Standards Alignment Guide**—How do you know which mathematical practices to incorporate in a lesson based on a given mathematics topic? Key content words are listed that align with suggested mathematics practice standards.

- **Content and Practices Alignment Protocol**—Teachers have an opportunity to make purposeful connections between the content standards and the mathematical practice standards by considering the thinking skills and content targeted in a particular standard.
- **Math Practice Prompts and Questions**—Once a math practice is selected, teachers must use instructional strategies to help students develop these dispositions. The use of iconic prompts and explicit questioning strategies are addressed in this section.

Understanding the Content Standards

Math content standards define what students should know and be able to do in the discipline of mathematics. They serve as objective benchmarks that align instruction and curriculum. These standards have been internationally benchmarked. Student performance is defined in terms of progress made toward the understanding of the content standards. The Core Knowledge Foundation (1999) suggests that standards represent a core of shared knowledge that students should learn in schools. In understanding this core knowledge embodied by the Common Core State Standards, it is important to first understand how they are structured.

The Common Core Content Standards are arranged by grade level expectations, with the exception of conceptual category designation at the high school level. As shown in Figure 1.1, domains exist inside each grade level set, clusters exist within domains, and individual standards exist within clusters.

Figure 1.1 Grade Level Standards

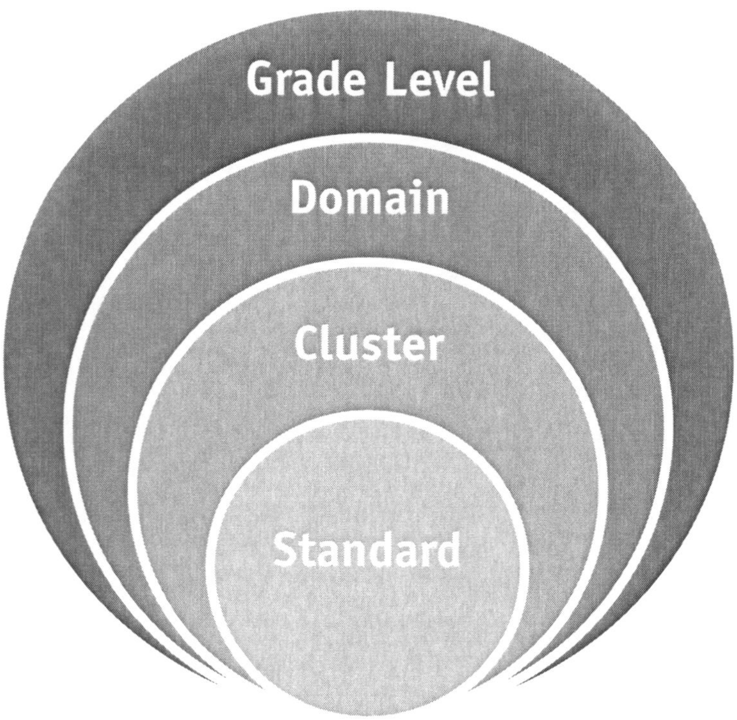

Grade level refers to students' year in school.

Domains are large groups of related standards that often extend across grade levels.

Clusters summarize groups of closely related standards.

Standards define what students know and should be able to do.

Reading the Content Standards

The standards for kindergarten through grade 8 are organized by grade level and clusters. For example, 3.NBT signals grade level 3 Number and Operations in Base Ten domain. Specific standards are listed within each domain and cluster, such as 3.NBT.2 (the second standard within grade level 3 Number and Operations in Base Ten domain). The standards can be written with and without cluster designation. When cluster designation is used, the insertion of capital letters (A, B, C, etc.) following the domain represent the sequence of the clusters in each domain. The following examples show this difference.

Without cluster designation:

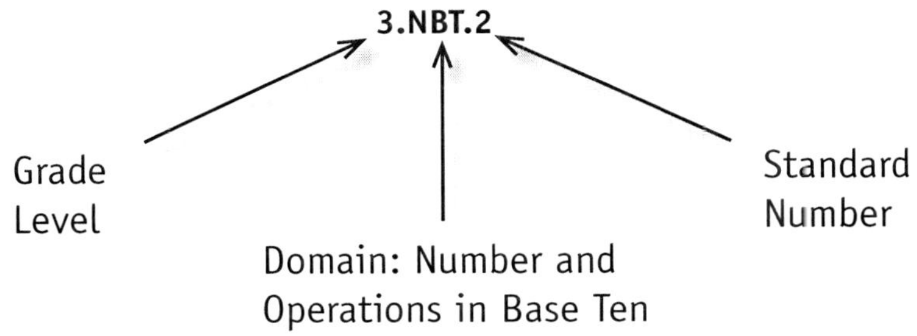

With cluster designation:

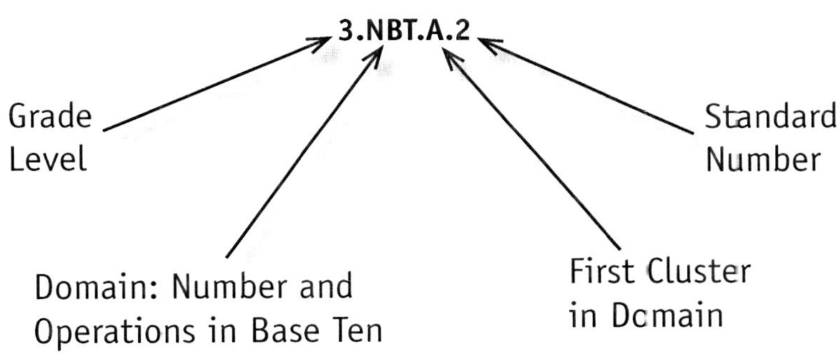

Figure 1.2 shows elements found within the third-grade level of the Common Core State Standards–Math.

Figure 1.2 Elements of the CCSS–M Grade Level Standards

Domain

Number and Operations in Base Ten **3.NBT**

Standard

1. Use place value understanding to round whole numbers to the nearest 10 or 100.
2. Fluently add and subtract within 1000 using strategies and algorithms based on place value, properties of operations, and/or the relationship between addition and subtraction.
3. Multiply one-digit whole numbers by multiples of 10 in the range 10-90 (e.g., 9 x 80, 5 x 60) using strategies based on place value and properties of operations.

Cluster

(CCSSO, NGA 2010a, 2,5)

Grade-Level Domains K-8

Domains represent large groups of clusters and inclusive standards. Figure 1.3 shows which domains are covered in each grade level and the progression of domains across grade levels. The common shading represents a learning trajectory of connected content from kindergarten through the conceptual categories represented in high school. The color-coordinated progressions show how mathematical structures maintain and develop over grade levels. This also provides an opportunity for the teacher to see how the standards they teach in their unique grade level relate to other standards across grade levels. Figure 1.3 can be found in full color on the Digital Resource CD.

Figure 1.3 Common Core Domain Coherence

<table>
<tr><th>Kindergarten</th><th>1</th><th>2</th><th>3</th><th>4</th><th>5</th><th>6</th><th>7</th><th>8</th><th>HS</th></tr>
<tr><td>Counting and Cardinality</td><td colspan="8"></td><td rowspan="3">Number and Quantity</td></tr>
<tr><td colspan="6">Number and Operations in Base Ten</td><td colspan="2">Ratios and Proportional Relationships</td><td></td></tr>
<tr><td colspan="3"></td><td colspan="3">Number and Operations—Fractions</td><td colspan="3">The Number System</td></tr>
<tr><td colspan="6" rowspan="2">Operations and Algebraic Thinking</td><td colspan="3">Expressions and Equations</td><td>Algebra</td></tr>
<tr><td colspan="2"></td><td>Functions</td><td>Functions</td></tr>
<tr><td colspan="6">Geometry</td><td colspan="3">Geometry</td><td>Geometry</td></tr>
<tr><td colspan="6">Measurement and Data</td><td colspan="3">Statistics and Probability</td><td>Statistics and Probability</td></tr>
</table>

(Ohio Department of Education 2015)

Analyzing the Content Standards

Understanding how to read the standards from an organizational standpoint is only the beginning of the analysis process. Meaningful understanding and application of the standards comes from an interrogative analysis that includes deconstructing them. Knowledge is contested and constructed. Several tools follow that aid in this process. However, analysis of the Common Core State Standards-Mathematics (CCSS–M) is twofold. In this chapter, you will find tools for analyzing the content standards and resources for analyzing the standards for mathematical practice. Both sets of standards should be analyzed with equal intensity.

Content Standard Analysis Protocol

The Content Standard Analysis Protocol helps teams of teachers, or individual teachers, closely examine the content standards. After this initial examination, the analysis protocol can be used routinely to explore and increase understanding of the standards. This is ideal to use as an ongoing measure to incorporate the standards into your practice.

Step 1: Make a copy of the Content Standard Analysis Chart for your grade level. Blank templates with the necessary cluster information for each grade level K–8 are on the Digital Resource CD. Download a copy of the CCSS from www.corestandards.org. Make sure to use your own state's standards, as the standards vary slightly from state to state. Complete the chart by writing down the standards that answer the following questions.

- *Which standards in the cluster are familiar to me?*
- *Which standards are unclear to me?*
- *What questions do I have?*

Step 2: Compare your chart with charts made by other teachers on your grade-level team. Discuss the procedural and conceptual application of the standards.

Step 3: Collaborate with your grade-level team to develop a better understanding of the standards. Spend time as a group asking questions and sharing answers. Use the Comprehensible Content Knowledge section on page 26 to reach consensus about any areas of ambiguity. You may find that you and your colleagues need to brush up on new or unfamiliar concepts that are part of your grade-level standards. It is important to take the time to acquire these skills so you feel confident teaching them to your students. The subsequent chart provides a partial example to help you get started.

Figure 1.4 Sample Content Standard Analysis Chart

Content Standard Analysis Chart Grade 5			
Cluster	**Which standards in the cluster are familiar to me?**	**Which standards are unclear to me?**	**What questions do I have?**
Operations and Algebraic Thinking (5.OA)			
Write and interpret numerical expressions.	*5.OA.1* *5.OA.2*		*How might I determine if parentheses and/or brackets are needed in the expression?*
Analyze patterns and relationships.		*5.OA.3*	*The standard says to generate patterns from a given rule. Am I to assume that the reverse is also expected? A pattern is provided and students are to generate a rule?*
Number and Operations in Base Ten (5.NBT)			
Understand the place value system.	*5.NBT.2* *5.NBT.4*	*5.NBT.1*	*How might I write a decimal in expanded form?*

You Try It!

1. Download the Content Standard Analysis Chart for your grade level from the Digital Resource CD (filename: contentanalysis.pdf).
2. Review the Content Standard Analysis Protocol (page 22).
3. Follow the protocol to complete the chart.
4. Decide on a plan to find answers to your remaining questions. For example, talk with grade-level peers, take an online learning course, or select some books or articles for professional growth.

Cluster Analysis Protocol

This analysis protocol can be used to understand the detailed meaning of each standard and its relationship to the overarching clusters. You will have an opportunity to explain each mathematical concept (written and illustrative) as well as identify the targeted thinking skill (level of cognitive demand) inherent in each standard. Key vocabulary is also highlighted.

Step 1: Make a copy of the Cluster Analysis Chart for each cluster in your grade level. A blank template is provided on the Digital Resource CD.

Step 2: Identify the following:

- How are the standards in the cluster related? How do they extend from each other?
- What are the overarching mathematical concepts represented? Illustrate these concepts symbolically or pictorially.
- What are the targeted thinking skills used to negotiate the content? What are examples of the application of the thinking skills?

Step 3: Determine what additional professional development is needed to increase your content knowledge.

The subsequent chart provides a partial example to help you get started.

Figure 1.5 Sample Cluster Analysis Chart

Grade Level

5th

Domain	**Cluster**
Operations & Algebraic Thinking	*Write and interpret numerical expressions.*

Standards

5.O.A.1 *5.OA.2*

Describe and illustrate the mathematical concepts being addressed.

Students work with expressions involving multiple operations to determine how to use ordering symbols.

Example 1: $(16 + 12) \div 4$

Example 2: Interpret. $3 \times (25 + 4)$

$(25 + 4) + (25 + 4) + (25 + 4)$ → *3 groups of* $(25 + 4)$

OR

3 groups of 25 plus 3 groups of 4

What are the thinking skills targeted? (evaluate, apply, identify, compare, or prove) What are examples of the application of the thinking skills?

use, evaluate, interpret

Example: Interpret the expression $5(170 + 63)$.

(What does the expression mean?)

How are the standards related to each other?

Standard 5.OA.1 uses symbols to evaluate a given expression.

Standard 5.OA.2 requires the student to create the expression and determine its meaning.

Key Vocabulary

- *numerical expressions*
- *parentheses*
- *brackets*

You Try It!

1. Download the Cluster Analysis Chart from the Digital Resource CD (filename: cluster.pdf).
2. Review the Cluster Analysis Protocol (page 24).
3. Follow the protocol to complete the chart.
4. Collaborate with your grade-level colleagues to share your ideas, and consider additional opportunities to increase your content knowledge.

Comprehensible Content Knowledge

A teacher's ability to help students understand the Common Core State Standards is highly dependent upon the teacher's own command of that content knowledge. Research shows that teachers must have a strong command of math content knowledge in order to be effective teachers (Tatto and Senk 2011). A teacher's increase in mathematical content knowledge has a positive impact on student performance (Hill et al., 2005). Since it is not always possible to take university courses, school districts ideally need to offer professional development time that includes content knowledge development.

However, if your district does not provide this type of professional development or you want to supplement your content knowledge, several state departments of education have created mathematics frameworks. These frameworks present the Common Core content in a comprehensible manner. They include content analysis as well as instructional and curricular strategies. After using the analysis protocols to identify areas of confusion, these frameworks can be used to improve your content knowledge.

The following are websites that can support you as you seek to increase your content knowledge:

Achieve the Core

www.achievethecore.org

Arizona Department of Education

http://www.azed.gov/azccrs/mathstandards/

California Department of Education

www.cde.ca.gov/ci/ma/cf/

Illustrative Mathematics

www.illustrativemathematics.org

Inside Mathematics

www.insidemathematics.org

National Council of Teachers of Mathematics

www.nctm.org

North Carolina State Board of Education

http://www.ncpublicschools.org/acre/standards/common-core-tools/

Understanding the Standards for Mathematical Practice

Alongside the content standards exist the standards for mathematical practice. These interdependent standards are inextricably linked and should not be taught in isolation. The standards for mathematical practice define the actions and dispositions of mathematically proficient students. These are expectations for all students in grades K–12. In all actuality, they extend through life as students continue to understand the world through a mathematical lens beyond high school. The standards for mathematical practice are as follows:

Standards for Mathematical Practice

1. Make sense of problems and persevere in solving them.
2. Reason abstractly and quantitatively.
3. Construct viable arguments and critique the reasoning of others.
4. Model with mathematics.
5. Use appropriate tools strategically.
6. Attend to precision.
7. Look for and make use of structure.
8. Look for and express regularity in repeated reasoning.

Figure 1.6 provides a description for the standards for mathematical practice and ways in which students can demonstrate their proficiency of each standard.

Figure 1.6 Standards for Mathematical Practice Proficiency Guide

Standard for Mathematical Practice	Description	How Students Demonstrate Proficiency
Make sense of problems and persevere in solving them.	There are two components of this practice: 1. Making Sense—Understanding what the problem is asking and the relationship between the parts of the problem. 2. Persevering—Having the resolve to stay with the problem through multiple attempts while being comfortable with disequilibrium that may be caused.	• working through multiple attempts • verifying that the solution makes sense in relationship to the problem • determining what is known and unknown in the problem • analyzing inherent constraints

Standard for Mathematical Practice	Description	How Students Demonstrate Proficiency
Reason abstractly and quantitatively.	There are two components of this practice: 1. Reasoning abstractly—Understanding how to represent a math context symbolically or algebraically (decontextualize) and how to represent a mathematical sentence in a particular real-world context. 2. Reasoning quantitatively—Understanding that the numbers involved represent specific quantities or amounts.	• converting math situations into symbols • converting math numbers and symbols into real-world situations • interpreting the value of a number • describing characteristics of numbers (whole, rational, etc.)
Construct viable arguments and critique the reasoning of others.	There are two components of this practice: 1. Constructing—Understanding how to establish a position or claim and support it with evidence, justifying the connections between the evidence and the claim. 2. Critiquing—Understanding how to judge against established criteria, noting ambiguities, providing counterexamples, and noting assumptions in arguments presented.	• providing explanation in conjunction with an "answer" • generating questions for an argument presented • providing counterexamples for a particular case • providing qualifying statements (e.g., exceptions for different cases) • justifying a proof (solution pathway) with mathematical rules
Model with mathematics.	Understanding how to not only represent a real-world situation with a mathematical model, but also how to test the model through repeated cycles of revision to find an accurate representation of the scenario. The model and the process of testing are then reported.	• contextualizing and decontextualizing a math scenario • interpreting the results of the math model in relationship to the real-world scenario • revising the math model through experimental testing • presenting an argument to defend the model and the modeling process conducted

Standard for Mathematical Practice	Description	How Students Demonstrate Proficiency
Use appropriate tools strategically.	Understanding what tools and resources are needed to solve a problem. Understanding the procedures that are inherent in the application of the tool.	• selecting math tools based on interpretation of the problem • explaining multiple resources that can be used to solve the problem • justifying selection of a tool based on the context of the problem
Attend to precision.	Understanding the appropriate use of academic vocabulary and labeling of quantities and objects within a math problem. Attention is given to all facets of the problem.	• using the proper math terminology in discussions • labeling units of measurement, axes, and other terms to remove ambiguity • using multiple calculations to check a derived answer
Look for and make use of structure.	An understanding exists regarding the relationships between the parts of a particular system and the impact of these parts on each other and the whole.	• identifying components within a problem and applying this information to solve the problem • describing the impact of changing a particular part of a problem on other parts of the problem • classifying a particular problem based on its features
Look for and express regularity in repeated reasoning.	Searching for repetition in problems and using this information in order to make general rules. Forecasting based on noted commonalities. Deriving patterns that define a particular problem.	• highlighting patterns and making predictions based on the pattern • creating generalizations for future situations that have similar constraints • recalling repeated results in order to make determinations

How the Practice Standards Relate

The standards for mathematical practice represent a framework for mathematical thought. These interdependent standards can be clustered showing the relationships that exist. A primary author of the Common Core State Standards for Mathematics, McCallum (2011), provides an organization of the standards. The eight standards are individually numbered and grouped based on relatedness. MP1 and MP6 are considered to be overarching standards. This is due in part to the general connection to all mathematical content. Being precise and understanding the context of the problem is essential in problem negotiation. MP2 and MP3 can be grouped because they both emphasize mathematical reasoning and explanation of the problem. MP4 and MP5 can be grouped because they describe modeling applications and the tools needed to create those models. MP7 and MP8 are often considered as the lifeline of mathematics—the ability to generalize based on an understanding of structure.

> To learn more about the relatedness of the mathematical practice standards, visit *www.pdesas.org/main/fileview/M3grouping_mathematical_practices.pdf.*

Figure 1.7 Grouping the Practice Standards

1. Make sense of problems and persevere in solving them 6. Attend to precision	2. Reason abstractly and quantitatively 3. Construct viable arguments and critique the reasoning of others	Reasoning and explaining
	4. Model with mathematics 5. Use appropriate tools strategically	Modeling and using tools
	7. Look for and make use of structure 8. Look for and express regularity in repeated reasoning	Seeing structure and generalizing

(McCallum 2011)

Standards for Mathematical Practice Analysis Protocol

A comprehensive analysis of the mathematical practices should include an examination of what each standard means, how the standards relate to one another, and how they align with the content standards. The following protocols provide an opportunity for this analysis.

The Standards for Mathematical Practice Analysis Protocol helps teachers and teams of teachers closely examine the meaning of the mathematical practices and how they are connected. Teachers are able to understand the connections that exist between the math practices. This is ideal to use as a first step in incorporating the Common Core mathematical standards into your practice.

Step 1: Make a copy of the Standards for Mathematical Practice Analysis Tool template. A blank template is provided on the Digital Resource CD.

Step 2: Use the standards for mathematical practice (www.corestandards.org) and the descriptions found in Figure 1.6 to complete the analysis tool that follows. Provide detailed descriptions of how this is taught in the context of your grade-level mathematics content.

The following example for Mathematical Practice 6 can be used to help you get started.

Figure 1.8 Sample Standards for Mathematical Practice Analysis Tool

CCSS Standards for Mathematical Practice 6: Attend to precision.

What does this standard mean?

Students use the proper vocabulary to name terms and concepts. Units of measurement are labeled. Clear definitions are used in explanations. Appropriate axes are labeled. Answers are compared to the question to determine if all aspects have been satisfied.

How is this standard connected to Math Practice 1?

An aspect of MP 1 is to make sense of problems. In order to make sense, terms have to be identified and units of measurement have to be considered. Unit conversions may be necessary.

Provide an example of the practice standard applied at your grade level and an explanation of the application.

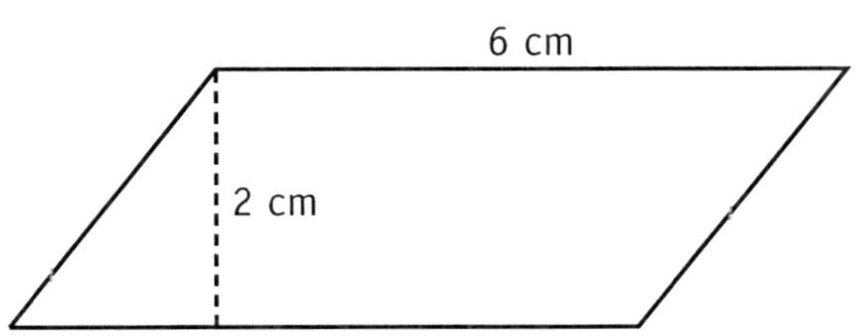

Area $= 2 \times 6 = 12$

Precision: label units of measurement appropriately

$A = 2 \text{ cm} \times 6 \text{ cm} = 12 \text{ cm}^2$

Chapter 1

You Try It!

1. Download the Standards for Mathematical Practice Analysis Tool template from the Digital Resource CD (filename: analysis.pdf).
2. Review the Standards for Mathematical Practice Analysis Protocol (page 31).
3. Follow the protocol to complete the chart.
4. Share your ideas, examples, and explanations with your grade-level peers. Discuss how the Standards for Mathematical Practice are connected.

Aligning Content and Practice Standards

An argument can be made that connects any math practice with any content standard. However, this random approach to alignment may not accentuate the salient points within a given content standard. Standards for Math Practice 1 and 6 represent overarching habits of mind that should always be present. However, specific alignment can be made by examining the nature of the content standard and cluster. The following chart includes key words/concepts within the content standards/clusters that can help you strategically align these standards to the practice standards.

Figure 1.9 Standards Alignment Guide

Standard for Mathematical Practice	Key Words/Concepts in the Standard/Cluster	
Make sense of problems and persevere in solving them.	• understand • describe	• determine known and unknown values • explain
Reason abstractly and quantitatively.	• determine • decontextualize or contextualize • calculate • relate number types	• create math sentences • compare numbers • illustrate • prove
Construct viable arguments and critique the reasoning of others.	• explain your reasoning • justify • defend a position	• validate or invalidate a point • prove
Model with mathematics.	• solve real-world problems • interpret results • create a representative model	• verify results • test a model
Use appropriate tools strategically.	• measure • estimate • illustrate	• construct • apply technology
Attend to precision.	• use appropriate units of measurement • label parts	• use academic vocabulary • answer all components of a problem
Look for and make use of structure.	• determine operations • construct factors • understand relationship between parts	• classify • compare items • examine equations/ expressions
Look for and express regularity in repeated reasoning.	• determine patterns • create rules	• predict a sequence • forecast

Content and Practices Alignment Protocol

Strategically aligning the content and practice standards is essential when designing Common Core lessons. The protocol that follows can assist you in aligning the content and practice standards.

Step 1: Make a copy of the Content and Practices Alignment Tool template. A blank template is provided on the Digital Resource CD.

Step 2: Identify the content standard(s) and cluster to be addressed in a given lesson.

Step 3: Identify the targeted thinking skills, such as Bloom's Taxonomy or critical-thinking skills.

Step 4: Identify the mathematical concept(s) within the content standard(s).

Step 5: Identify one or two specific math practices that best align with the content standards. Use the Standards Alignment Guide (Figure 1.9) for support.

Step 6: Record your rationale for the selected math practices.

An example of this strategic alignment is provided in Figure 1.10 to help you get started.

Figure 1.10 Sample Content and Practices Alignment Tool

Domain: *Operations and Algebraic Thinking*	
Cluster: *Represent and solve problems involving multiplication and division.*	
Standard(s): *3.OA.4 Determine the unknown whole number in a multiplication or division equation relating three whole numbers.*	
Thinking Skill(s): *determine, relating*	**Mathematical Concept(s):** *Unknown whole numbers; multiplication and division (inverse operations)*
Math Practice(s): *MP2 and MP7* **Rationale:** *Math Practice Standard 2: Reason abstractly and quantitatively.* *The standard targets determining unknown whole numbers—This involves quantitative reasoning which includes understanding number types and characteristics of certain number classifications (e.g., whole numbers, rational numbers, irrational numbers).* *Math Practice Standard 7: Look for and make use of structure.* *The standard also targets relating inverse operations. This emphasizes the idea of understanding the relationship between the opposite operations of multiplication and division. The concept of structure emphasizes how parts (operations, in this example) impact each other. This standard also addresses the concept of an equation, which has specific structured parts (e.g., equal sign, expressions).*	

You Try It!

1. Download the Content and Practices Alignment Tool template from the Digital Resource CD (filename: alignment.pdf).

2. Review the Content and Practices Alignment Protocol (page 34).

3. Follow the protocol to complete the chart.

4. Share your ideas and rationale for your selected practice standards with your grade-level colleagues. Determine whether additional practice standards can be aligned with the same content standard.

Using Prompts and Questions to Support Student Learning

The use of visual prompts may help students become intentionally aware of the standards for mathematical practice. It is important to discuss the purpose of using prompts with the class. Prompts are used to alert students to a particular practice. Teachers can prominently display these prompts and provide a general overview of each mathematical practice and the corresponding prompt. As teachers and students negotiate math problems, specific prompts can be highlighted by posting the prompt next to the problem. Posting these prompts and discussing how they are used makes the tacit curriculum explicit. After repeated exposure, students become more aware of the mathematical practices and eventually use them in a self-regulated manner. This includes self-selection of particular prompts based on a given math context and the student's justification for their selection. Figure 1.11 represents a series of prompts.

There are times when the math practices may be superordinate in the context of a lesson. For example, if you notice that students are able to develop a mathematical model that represents a given concept but are unable to refine this model through repeated testing, you may want to highlight the process of modeling as the major concept in the lesson. The students would derive an answer but this would be subordinate to a discussion regarding the concept of modeling. This discussion would be used to understand a student's ability to develop conjectures, test these conjectures, and determine the validity of their answers (MP4). Making the mathematical practice central in a lesson should be done periodically to emphasize the importance of the mathematical practices.

Enlarged versions of these visual prompts in color are included on the Digital Resource CD.

Figure 1.11 Visual Prompts

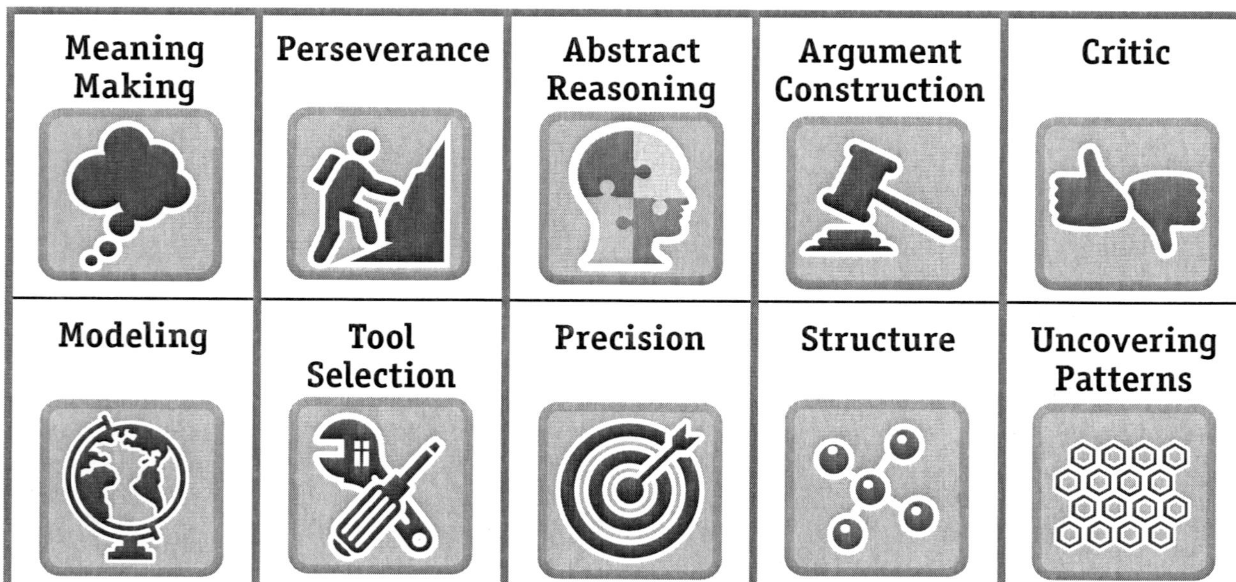

Once the mathematical practices have been selected for a particular task and prompts are visually displayed, questioning strategies by both teachers and students can be used to develop the mathematical practice. Encourage students to generate questions beyond those provided in Figure 1.12.

Figure 1.12 Questions and Student Actions for the Mathematical Practice (MP) Standards

Student Actions	Teacher-Directed Questions
MP1: Make sense of problems and persevere in solving them.	
• Interpret and make meaning of the problem to find a starting point. Analyze what is given in order to explain the meaning of the problem. • Plan a solution pathway instead of jumping to a solution. • Monitor student progress and change the approach if necessary. • See relationships between various representations. • Relate current situations to concepts or skills previously learned and connect mathematical ideas to one another. • Continually ask, "Does this make sense?" • Understand various approaches to solutions.	• How would you describe the problem in your own words? • How would you describe what you are trying to find? • What do you notice about ________? • What information is given in the problem? • What is unknown in the problem? • How might you describe the relationship between the quantities? • Describe what you have already tried. What might you change? • What are some other strategies you might try? • What are some other problems that are similar to this one? • How might you use one of your previous problems to help you begin? • How else might you organize/represent/show ________?
MP2: Reason abstractly and quantitatively.	
• Make sense of quantities and their relationships. • Decontextualize (represent a situation symbolically and manipulate the symbols) and contextualize (make meaning of the symbols in a problem) quantitative relationships. • Understand the meaning of quantities and remain flexible in the use of operations and their properties. • Create a logical representation of the problem. • Attend to the meaning of quantities, not just how to compute them.	• What do the numbers used in the problem represent? • What is the relationship of the quantities? • How is ________ related to ________? • What is the relationship between ________ and ________? • What does ________ mean to you? (*symbol, quantity, diagram*) • What properties might we use to find a solution? • How can this be represented symbolically, pictorially, or in words? • How can you express this answer in the context of the problem?

<table>
<tr><th>Student Actions</th><th>Teacher-Directed Questions</th></tr>
<tr><td colspan="2">MP3: Construct viable arguments and critique the reasoning of others.</td></tr>
<tr><td>• Analyze problems and use stated mathematical assumptions, definitions, and established results in constructing arguments.
• Justify conclusions with mathematical ideas.
• Listen to the arguments of others and ask useful questions to determine if an argument makes sense.
• Ask clarifying questions or suggest ideas to improve/revise the argument.
• Compare two arguments and determine correct or flawed logic.</td><td>• What mathematical evidence would support your solution?
• How can we be sure that ________?
• How could you prove that ________?
• Will it still work if ________?
• What were you considering when ________?
• How did you decide to try that strategy?
• How did you test whether your approach worked?
• What assumptions were made in the argument?
• How did you decide what the problem was asking you to find?
• Did you try a method that did not work?
• What is ambiguous in the argument presented?
• What is the same and what is different about ________?
• How could you demonstrate a counterexample?</td></tr>
<tr><td colspan="2">MP4: Model with mathematics.</td></tr>
<tr><td>• Understand this is a way to reason quantitatively and abstractly (able to decontextualize and contextualize).
• Apply known mathematics to solve everyday problems.
• Simplify a complex problem and identify important quantities to look at relationships.
• Represent mathematics to describe a situation with either an equation or a diagram and interpret the results of a mathematical situation.
• Reflect on whether the results make sense, possibly improving/revising the model.
• Ask, "How can I represent this mathematically?"</td><td>• What questions can I ask of this real-world scenario?
• What equation could represent the problem?
• Does this model accurately represent this real-world situation?
• Have I validated the model through multiple tries?
• How will I communicate my findings to the public?
• What are some ways to represent the quantities?
• What is an equation or expression that matches the diagram/number line/chart/table?
• Where do you see one of the quantities from the task in your equation or expression?
• How would it help to create a diagram/graph/table?
• What are some ways to visually represent ________?
• What formula might apply in this situation?</td></tr>
</table>

Student Actions	Teacher-Directed Questions
MP5: Use appropriate tools strategically.	
• Use available tools to recognize the strengths and limitations of each. • Use estimation and other mathematical knowledge to detect possible errors. • Identify relevant external mathematical resources to pose and solve problems. • Use technological tools to deepen mathematical understanding.	• What mathematical tools can be used to visualize and represent the situation? • What resources are more efficient to help solve the problem? • What approach are you considering trying first? • What estimate did you make for the solution? • In this situation would it be helpful to use a graph/number line/ruler/diagram/calculator/manipulative? • Why was it helpful to use ________? • What can using a ________ show us that using a ________ may not?
MP6: Attend to precision.	
• Communicate precisely and use clear mathematical language when discussing reasoning. • Understand the meanings of symbols used in mathematics and label quantities appropriately. • Express numerical answers with a degree of precision appropriate for the problem's context. • Calculate efficiently and accurately.	• What mathematical terms apply in this situation? • How did you know your solution was reasonable? • Explain how you might show that your solution answers the problem. • Have you answered all of the parts of the question? • Have you labeled the appropriate parts (axes, units)? • What symbols or mathematical notations are important in this problem? • What mathematical language, definitions, and properties can you use to explain ________? • How could you check your solution to see if it answers the question?

Student Actions	Teacher-Directed Questions
MP7: Look for and make use of structure.	
• Apply general mathematical rules to specific situations. • Look for the overall structure and patterns in mathematics. • See complex things as single objects or as being composed of several objects.	• What observations do you make about ________? • What do you notice when ________? • What parts of the problem might you eliminate/simplify? • What patterns do you find in ________? • How do you know if something is a pattern? • What previously learned ideas were useful in solving this problem? • How do the individual parts help you understand the entire problem? • How does this relate to ________? • In what ways does this problem connect to other mathematical concepts?
MP8: Look for and express regularity in repeated reasoning.	
• See repeated calculations and look for generalizations and shortcuts. • See the overall process of the problem and attend to the details. • Understand the broader application of patterns and see the structure in similar situations. • Continually evaluate the reasonableness of their intermediate results.	• How does this strategy work in other situations? • Is this always true, sometimes true, or never true? • How would you prove that ________? • What do you notice about ________? • What is happening in this situation? • What would happen if ________? • Is there a mathematical rule for ________? • What predictions or generalizations can this pattern support? • What mathematical consistencies do you notice?

Once the prompts have been selected based on an understanding of the targeted content, the teacher places the prompts prominently on the board. Associated questions drawn from the chart above are identified. As the teacher asks each question, each prompt is raised to create an association between the prompt and the desired student action.

The following two examples illustrate how the visual prompts and questions can be used in the classroom.

Scenario 1

(3rd grade CCSS: Operations & Algebraic Thinking—3.OA.2, 3.OA.8)

Two groups of students from Galiano Elementary School were walking to the library when it began to rain. The 7 students in Mr. Corallo's group shared 3 large umbrellas with Ms. Well's group of 11 students. If the same number of students were under each umbrella, how many students were under each umbrella?

Possible visual prompts:

Meaning Making

Abstract Reasoning

Argument Construction

Critic

What factors are known and what factors are unknown? (Meaning Making)

This targets MP1, specifically making meaning of problems. Students may answer the questions, but may be unaware of the mathematical skill being developed. As the teacher asks the students the question, the selected MP1 prompt can be displayed (example:). The teacher engages the students in conversation of what it means to make meaning of a problem and the importance of the skill.

What are the details of the problem? (Meaning Making)

Students may list the number of students present in each group and the number of available umbrellas, and the constraint of having the same number of students under each umbrella. Students may explain that the problem is asking them to find out how many students were under each umbrella.

Students may list the number of initial groups, the relevant details (e.g., teachers' names vs. numerical data), and the type of answer solicited.

How can this problem be written symbolically, numerically, or pictorially? (Abstract Reasoning)

Students may draw a picture similar to the following:

Students may write an algebraic equation similar to the following:

$3 \times n = 7 + 11$

What is your rationale? (Argument Construction)

Students may provide written responses detailing their selected method of problem solving, including an explanation of the solution and the supportive evidence.

How strong is the evidence used in the arguments presented? (Critic)

Students may question the responses presented by their classmates (e.g., similarities and differences between selected problem-solving approaches, additional considerations that may have been taken).

Scenario 2:

(8th grade CCSS: Expressions and Equations—8.EE.2)

Solve the following quadratic equation: $x^2 = 49$. Explain your rationale.

Possible visual prompts:

Meaning Making

Structure

Argument Construction

Patterns

Critic

What does the degree of the polynomial tell you about the root? (Meaning Making and Structure)

Students may respond that the exponent 2 represents the highest degree of the polynomial. Therefore, a square root may be used to solve the equation.

Discuss the possible solution methods for this quadratic equation. (Argument Construction and Patterns)

Students may choose to find the difference of two squares, use the square root method, or apply the quadratic formula.

Compare and contrast the methods presented. Which methods may be advantageous for this problem? Why? (Argument Construction and Critic)

Students may present the procedural applications for solving the quadratic equation using various methods. Students may present arguments for selected methods based on the structure of the polynomial.

Some students may have difficulty articulating their understanding when responding to these questions and prompts. The following sentence frames can provide access to the language of mathematical discourse and communication required by today's standards.

Mathematical Practice Standard 1

- The things that are known or given in the problem are ________.
- The things that are unknown in the problem are ________.
- This problem means ________ because of ________.
- The answer of ________ means that ________ in relation to the problem.

Mathematical Practice Standard 2

- The number ________ represents ________.
- I can draw a ________ to represent the ________ in the problem.
- This ________ is measured in units of ________.
- This ________ can be represented as ________ in real life.

Mathematical Practice Standard 3

- I believe the answer is ________ because ________.
- I used the method of ________ to solve the problem.
- My argument is ________ based on ________.
- I agree/disagree with ________ because of ________.
- The evidence of ________ supports the ________.
- The counterexample of ________ disproves the ________.
- The statement of ________ is true except when ________.

Mathematical Practice Standard 4

- The real-world context is ________.
- I can use ________ to represent the connection between ________.
- The model used is ________ and shows how ________.
- I believe this is a valid representation of ________ because ________.
- I tested this model by ________ and it shows that ________.

Mathematical Practice Standard 5

- We use ________ to measure ________.
- The tool of ________ is helpful when ________.
- The resources needed to solve include ________.

Mathematical Practice Standard 6

- The ________ is defined as ________.
- The ________ is the mathematical language for ________.
- The ________ is measured in ________ units.

Mathematical Practice Standard 7

- The ________ demonstrates the pattern ________ because ________.
- The ________ of the problem relates to ________.
- The ________ is composed of ________.

Mathematical Practice Standard 8

- I notice that ________ is repeated because ________.
- The formula of ________ can be used when ________.
- The pattern of ________ suggests that ________.
- I predict that ________ based on ________.

Questioning to Build Conceptual Understanding

It is not uncommon in the math classroom to hear students stating, *"I know how to do this."* While that may be true from a procedural standpoint, the standards can be used to examine and extend a student's claim of "knowing" a particular concept. This examination can occur by changing one word in the statement, "I know how to do this." Changing one word assesses multiple knowledge domains (conceptual, procedural, metacognition) and targets the development of the standards for mathematical practice. If we focus on the function words of *how*, *why*, *where*, and *when*, an interesting cognitive shift occurs that unveils a student's more comprehensive understanding of a concept.

- **Do you know *how* to do this?** Students that know *how* to do a particular concept will be able to represent that concept in multiple ways. They will be able to demonstrate concrete verbal and pictorial representations, numeric representations, and abstract symbolic representations. This demonstrates a student's application of Mathematical Practice 2. Students will be able to reason abstractly and quantitatively.

  Students that know *how* to do a particular concept will know what tools are needed to negotiate the task or scenario. Mathematical Practice 5 notes that students will be able to strategically choose tools that will aid in problem solving.

- **Do you know *why* to do this?** When students begin to understand the value of a mathematical concept, the likelihood that they will continue to use and retain understanding of that concept beyond the classroom increases. Students that know *why* are able to provide a rationale for the selection of an algorithm or law used to negotiate a problem. An explanation of *why* triggers the development of multiple math practices, including MP3 and MP1. Students will be able to make meaning of a problem and construct a viable argument as a result.

- **Do you know *where* to do this?** Mathematics is a discipline that describes patterns, relationships, and changes in everyday life. Mathematical Practice 4, modeling, captures the essence of why we study mathematics—to be able to interpret the world we live in. Students that know how to apply a math concept should be able to explain the context in life that calls for *where* that particular application is needed.

- **Do you know *when* to do this?** Students that know will be able to explain *when* a concept relates to other concepts and *when* that concept may be called upon in a logical progression of thought. One of the instructional shifts introduced by the standards is that of coherency. Knowing when a concept is needed and how it relates to other concepts being employed demonstrates a coherent understanding of the discipline. The ability to look for and make use of structure, generalize, and to know when to use these generalizations (Math Practices 7 and 8) demonstrates a comprehensive understanding of the mathematical concept.

Application Example 1 (7.G.4b)

Do you know ***how*** to find the area of the shape? If so, what is the area of this shape?

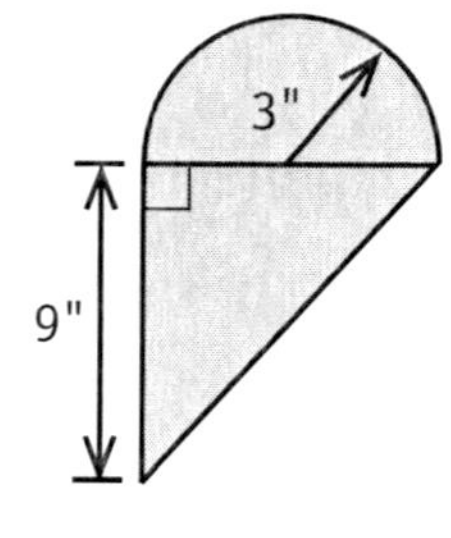

Circle	$A = \pi(3)^2 = 9\pi\ in.^2$
Semi-circle	$9\pi\ in.^2 \div 2 = 4.5\pi\ in.^2$
Triangle	$A = \frac{1}{2}(9)(6) = 27\ in.^2$
	Therefore,
Total Area =	$27 + 4.5\pi\ in.^2$

Do you know ***why*** you are applying this concept?

The shape presented is a composite of some more basic shapes, namely a triangle and a semi-circle. The formulas selected provide the area for a circle and a triangle. I divided the area of the circle in half to represent the total area of the semi-circle. In order to use the triangle formula, I needed to determine the value of the base and height. In a triangle, the height is the perpendicular line segment from any side (base) and the opposite vertex. Therefore, either 9 in. or 6 in. (the diameter of the circle) can serve as the base. I added the two individual areas to find the area of the composite shape. The value of π remains in the answer to demonstrate the precise value.

Do you know ***where*** to apply this concept?

This problem can be applied to solve any area problems in real life that involve the combination of many basic shapes, such as triangles, squares, rectangles, circles, etc. I can use this line of thought to calculate the total area needed for carpeting, painting, and landscaping specifications.

Do you know ***when*** to apply this concept?

I applied the area formula after I was able to determine needed information in the scenario such as the radius of the circle and type of triangle present (right, equilateral, etc.). Once it was determined that the base and height of the triangle were known, I was able to calculate the total area. This concept can be applied building upon the existing understanding of the areas of basic shapes and how these basic shapes can be combined to create composite shapes. Finding the areas of composite shapes leads to finding the area of irregular shapes not composed of basic shapes.

Application Example 2 (6.RP.3a)

Do you know ***how*** to solve this problem?

David has read a certain number of pages in a given time frame (see the table). How many pages do you project that David will read in 25 minutes?

Minutes	5	10	15	20	25
Pages	2	4	6	8	?

Based on the pattern that follows, David will have read 10 pages after 25 minutes.

2 × number of minutes = 5 × number of pages

Do you know ***why*** you are applying this concept?

There is a constant rate of change comparing the number of minutes reading to the number of pages read. For every change in time of five minutes there is a change in the number of pages by two. Recognizing this structure leads to the application of the concept.

Do you know ***where*** to apply this concept?

This concept can be applied when determining distance or time traveled after maintaining a constant speed. It can also be applied to calculate total savings based on a constant amount saved over a period of time. It can also be used to make scale models and drawings.

Do you know ***when*** to apply this concept?

After I determined that there is a consistent rate of change between minutes and pages, I was able to use this pattern to calculate the resulting number of pages. This concept relates to other concepts such as equivalent fractions and proportions.

As we continue to change the paradigm in mathematics education of what it means to be a student of mathematics, we will continue to hear the response of "I know how to do this." Redefining this idea of knowledge calls upon us as educators to ask the questions of not only "Do you know *how* to do this?", but also "Do you know *why* to do this?", "Do you know *where* to do this?", and "Do you know *when* to do this?" The precise (MP6) answers to these questions using the language of the discipline will help assess a student's claim of knowing and provide support, resulting in students developing as mathematicians.

Students can demonstrate their understanding of a mathematical concept by completing a Question Frame. Within the frame, students will show how to solve the particular concept in multiple ways, why the concept is applied, where the concept is applied in real life, and when the concept is applied.

A Question Frame can be used in multiple ways. A problem can be placed in the center of the frame. The student can then explore the four questions written in the parts of the frame and provide answers. The center section can also be cut out and the frame can be used as an overlay with existing curriculum. The topic of exploration can be centered in the middle of the frame. Students can also do this activity in cooperative groups by discussing the questions and taking turns being the recorder.

The Question Frame Template can be found on the Digital Resource CD.

Figure 1.13 Question Frame Template

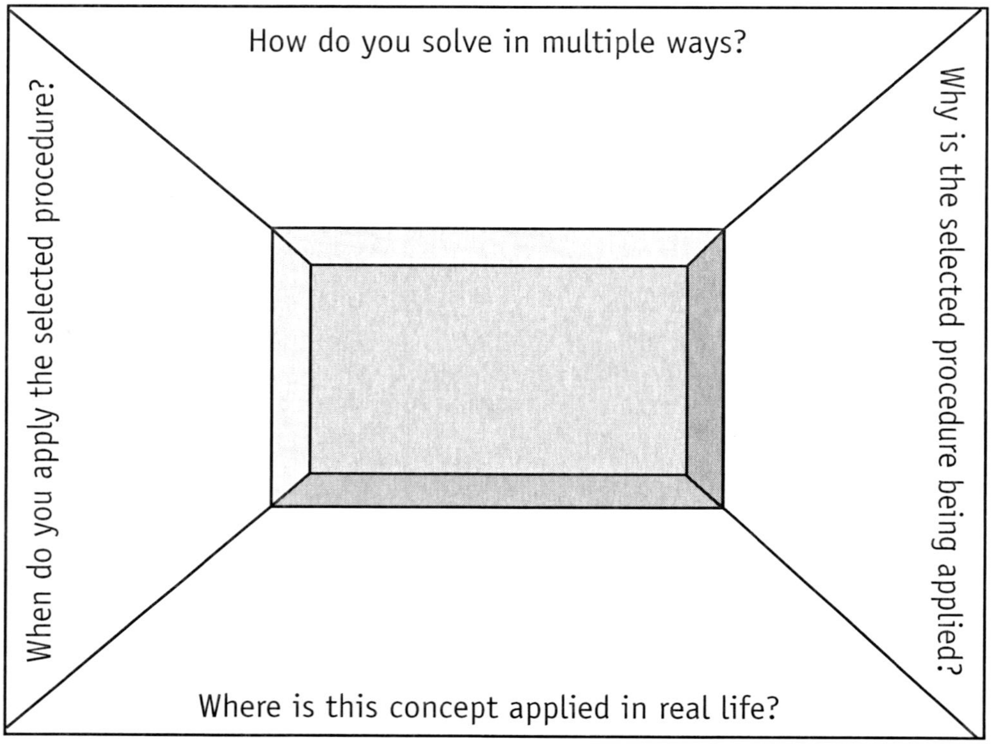

You Try It!

Think of a math concept that you know and have taught in your practice. Ask yourself the following questions:

1. Do you know *how* to apply this concept? (consider multiple representations, tools)
2. Do you know *why* you are applying this concept?
3. Do you know *where* to apply this concept?
4. Do you know *when* to apply this concept?
5. Do you have the evidence to validate your claim of knowing the concept?

What did you experience as a learner? (e.g., cognitive shift, anxiety)

What are the implications for your classroom instruction and curriculum design?

Stop and Think

Question 1 How has your knowledge of mathematics been reinforced through the use of the analysis protocols?

Question 2 How might your understanding of the mathematical practices impact your students' understanding of mathematics?

Question 3 How might you emphasize the mathematical practices within a given lesson?

Chapter 2

Instructional Shift of Focus

The instructional shift of focus outlines the importance of different grade-level topics—importance being defined as key understandings for the subsequent grade level and beyond. Focus requires teachers to significantly narrow and deepen the scope of the content addressed in the mathematics classroom. Time and energy should be spent on the concepts that are prioritized in the standards. Priority standards are high-leverage standards that have lasting implications for students' readiness for the next grade level as well as high school and university level mathematics. Prioritizing standards does not mean excluding standards. While all grade-level standards should be integrated into the curriculum and instructional delivery, they should not all be treated equally.

Focus has been approximately quantified based on preliminary blueprints from the Smarter Balanced Assessment Consortia (SBAC) and the Partnership for Assessment of Readiness for College and Careers (PARCC). Figure 2.1 shows approximations for these resource allocations.

Figure 2.1 Cluster Designations

70%	**Major Clusters:** Intensive areas of study to develop deep understanding
20%	**Supporting Clusters:** Designed to support and strengthen major areas' emphasis—rethinking and linking to major topics
10%	**Additional Clusters:** May not connect tightly to the other major work of the grade—expose to other subjects often across disciplines

The instructional shift of focus provides time for students to develop a comprehensive understanding of the topic and integrate this new information into their existing mathematics schema.

Focus and Rigor

Focus on the major clusters allows for a rigorous exploration of the content. Rigor includes a solid conceptual understanding, a high degree of procedural skill and fluency, and the ability to apply mathematics in a real-world context. Though the instructional shift of rigor will be addressed later in this book, it is important to note that the purpose of identifying major clusters is to ensure that they are explored in a rigorous fashion.

The instructional shift of focus also includes an understanding of what should *not* be taught at a particular grade level. This may require teachers to change their previous practices and commonly held beliefs regarding content taught at a particular grade level.

This chapter includes the following tools to target the shift of focus:

- **Content Emphases Charts**—This tool enables educators to provide a graphic illustration for themselves and the students. It outlines the key concepts in a grade level, fluency expectations, and cluster designations. These can be created as posters and used as a classroom guide.
- **Curriculum Focus Audit Tool**—This curriculum focus-auditing tool can be used to determine the degree of alignment with content that should be emphasized at each grade level.

A Closer Look at the Clusters

The Content Emphases Charts show the major, supporting, and additional clusters at each grade level. The major clusters are areas of intensive focus where students need to develop conceptual understanding, fluency, and application of the core concepts. These clusters require greater emphasis than the others based on the depth of the ideas, the time that they take to master, and/or their importance to future mathematics or the demands of college and career readiness. Seventy percent of instructional time should be spent on the major clusters. The supporting clusters are areas where some material is being covered but in a way that applies core understanding and supports and strengthens areas of major emphasis. Twenty percent of instructional time should be spent on the supporting clusters. The additional clusters expose students to other subjects and may not connect explicitly to the major work of the grade. Neglecting the content in the additional clusters will leave gaps in students' skills and understanding and will leave students unprepared for the challenges of the later grades. Ten percent of instructional time should be spent on the additional clusters. Figure 2.2 shows the breakdown of approximate time that should be spent during the year on each group of clusters.

Figure 2.2 Math Clusters

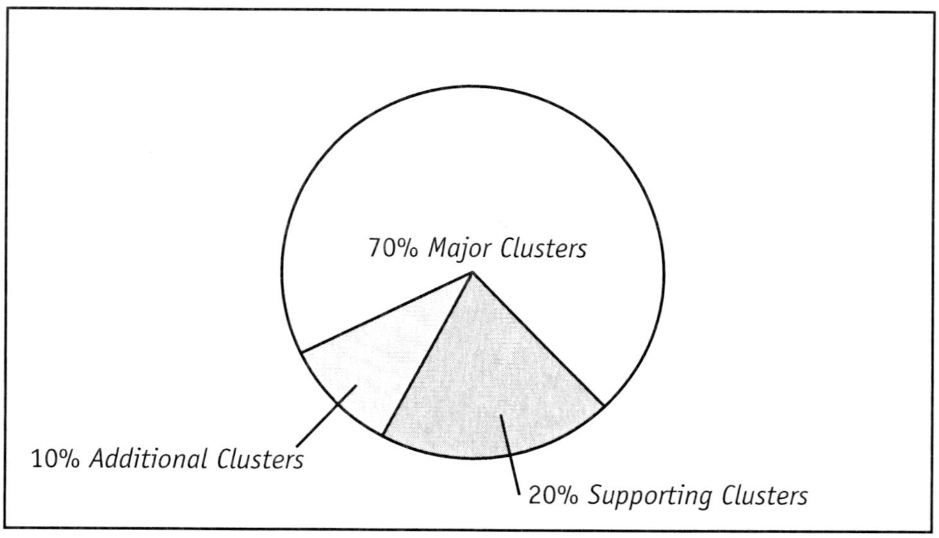

Using the Grade-Level Content Emphases Charts

The grade level content emphases charts can be enlarged and posted in the classroom to serve as visual reminders of what content is and is not addressed at a particular grade. These charts not only include content areas of focus but also expected procedural fluencies for each grade level.

Teachers can use these charts as a point of reference to determine curriculum focus alignment. The charts also serve as visual indicators for students regarding grade-level expectations.

Figure 2.3 Kindergarten Content Emphases

■—Major Cluster ●—Supporting Cluster ◆—Additional Cluster

Critical Focus Areas	
➡ Representing and comparing whole numbers, initially with sets of objects ➡ Describing shapes and space	
Expected Fluency	
➡ Add and subtract within 5.(K.OA.5)	
Domains and Clusters	
Counting and Cardinality	**K.CC**
■ Know number names and the count sequence. ■ Count to tell the number of objects. ■ Compare numbers.	
Operations and Algebraic Thinking	**K.OA**
■ Understand addition as putting together and adding to and understand subtraction as taking apart and taking from.	
Number and Operations in Base Ten	**K.NBT**
■ Work with numbers 11–19 to gain foundations for place value.	
Measurement and Data	**K.MD**
◆ Describe and compare measurable attributes. ● Classify objects and count the number of objects in categories.	
Geometry	**K.G**
◆ Identify and describe shapes. ● Analyze, compare, create, and compose shapes.	

Figure 2.4 First Grade Content Emphases

■—Major Cluster ●—Supporting Cluster ◆—Additional Cluster

Critical Focus Areas	
➡ Developing understanding of addition, subtraction, and strategies for addition and subtraction within 20	
➡ Developing understanding of whole number relationships and place value, including grouping in tens and ones	
➡ Developing understanding of linear measurement and measuring lengths as iterating length units	
➡ Reasoning about attributes of, and composing and decomposing geometric shapes	
Expected Fluency	
➡ Add and subtract within 10 (1.OA.6)	
Domains and Clusters	
Operations and Algebraic Thinking	**1.OA**
■ Represent and solve problems involving addition and subtraction.	
■ Understand and apply properties of operations and the relationship between addition and subtraction.	
■ Add and subtract within 20.	
■ Work with addition and subtraction equations.	
Number and Operations in Base Ten	**1.NBT**
■ Extend the counting sequence.	
■ Understand place value.	
■ Use place value understanding and properties of operations to add and subtract.	
Measurement and Data	**1.MD**
■ Measure lengths indirectly and by iterating length units.	
◆ Tell and write time.	
● Represent and interpret data.	
Geometry	**1.G**
◆ Reason with shapes and their attributes.	

Figure 2.5 Second Grade Content Emphases

■—Major Cluster ●—Supporting Cluster ◆—Additional Cluster

Critical Focus Areas	
➡ Extending understanding of base-ten notation. ➡ Building fluency with addition and subtraction. ➡ Using standard units of measure. ➡ Describing and analyzing shapes.	
Expected Fluency	
➡ Add and subtract within 20.(2.OA.2) ➡ Add and subtract within 100.(2.NBT.5)	
Domains and Clusters	
Operations and Algebraic Thinking	**2.OA**
■ Represent and solve problems involving addition and subtraction. ■ Add and subtract within 20. ● Work with equal groups of objects to gain foundations for multiplication.	
Number and Operations in Base Ten	**2.NBT**
■ Understand place value. ■ Use place value understanding and properties of operations to add and subtract.	
Measurement and Data	**2.MD**
■ Measure and estimate lengths in standard units. ■ Relate addition and subtraction to length. ● Work with time and money. ● Represent and interpret data.	
Geometry	**2.G**
◆ Reason with shapes and their attributes.	

Chapter 2

Figure 2.6 Third Grade Content Emphases

■—Major Cluster ●—Supporting Cluster ◆—Additional Cluster

Critical Focus Areas	
➡ Developing understanding of multiplication and division and strategies for multiplication and division within 100.	
➡ Developing understanding of fractions, especially unit fractions (fractions with numerator 1).	
➡ Developing understanding of the structure of rectangular arrays and of area.	
➡ Describing and analyzing two-dimensional shapes.	
Expected Fluency	
➡ Multiply and divide within 100 (3.OA.7)	
➡ Add and subtract within 1000 (3.NBT.2)	
Domains and Clusters	
Operations and Algebraic Thinking	**3.OA**
■ Represent and solve problems involving multiplication and division.	
■ Understand properties of multiplication and the relationship between multiplication and division.	
■ Multiply and divide within 100.	
■ Solve problems involving the four operations and identify and explain patterns in arithmetic.	
Number and Operations in Base Ten	**3.NBT**
◆ Use place value understanding and properties of operations to perform multi-digit arithmetic.	
Number and Operations-Fractions	**3.MD**
■ Develop understanding of fractions as numbers.	
Measurement and Data	**3.MD**
■ Solve problems involving measurement and estimation of intervals of time, liquid volumes, and masses of objects.	
● Represent and interpret data.	
■ Geometric measurement: understand concepts of area and relate area to multiplication and to addition.	
◆ Geometric measurement: recognize perimeter as an attribute of plane figures and distinguish between linear and area measures.	
Geometry	**3.G**
● Reason with shapes and their attributes.	

Figure 2.7 Fourth Grade Content Emphases

■—Major Cluster ●—Supporting Cluster ◆—Additional Cluster

Critical Focus Areas

- ➡ Developing understanding and fluency with multi-digit multiplication and developing understanding of dividing to find quotients involving multi-digit dividends.
- ➡ Developing an understanding of fraction equivalence, addition and subtraction of fractions with like denominators, and multiplication of fractions by whole numbers.
- ➡ Understanding that geometric figures can be analyzed and classified based on their properties, such as having parallel sides, perpendicular sides, particular angle measures, and symmetry.

Expected Fluency

- ➡ Add and subtract within 1,000,000. (4.NBT.4)

Domains and Clusters

Operations and Algebraic Thinking 4.OA

- ■ Use the four operations with whole numbers to solve problems.
- ● Gain familiarity with factors and multiples.
- ◆ Generate and analyze patterns.

Number and Operations in Base Ten 4.NBT

- ■ Generalize place value understanding for multi-digit whole numbers.
- ■ Use place value understanding and properties to perform multi-digit arithmetic.

Number and Operations–Fractions 4.NF

- ■ Extend understanding of fraction equivalence and ordering.
- ■ Build fractions from unit fractions by applying and extending previous understandings of operations on whole numbers.
- ■ Understand decimal notation for fractions, and compare decimal fractions.

Measurement and Data 4.MD

- ● Solve problems involving measurement and conversion of measurements from a larger unit to a smaller unit.
- ● Represent and interpret data.
- ◆ Geometric measurement: understand concepts of angle and measure angles.

Geometry 4.G

- ◆ Draw and identify lines and angles, and classify shapes by properties of their lines and angles.

Figure 2.8 Fifth Grade Content Emphases

■—Major Cluster ●—Supporting Cluster ◆—Additional Cluster

Critical Focus Areas	
➡ Developing fluency with addition and subtraction of fractions and developing understanding of the multiplication of fractions and of division of fractions in limited cases (unit fractions divided by whole numbers and whole numbers divided by unit fractions). ➡ Extending division to 2-digit divisors, integrating decimal fractions into the place value system, developing understanding of operations with decimals to hundredths, and developing fluency with whole number and decimal operations. ➡ Developing understanding of volume.	
Expected Fluency	
➡ Multiply multi-digit numbers. (5.NBT.5)	
Domains and Clusters	
Operations and Algebraic Thinking	**5.OA**
◆ Write and interpret numerical expressions. ◆ Analyze patterns and relationships.	
Number and Operations in Base Ten	**5.NBT**
■ Understand the place value system. ■ Perform operations with multi-digit whole numbers and with decimals to the hundredths.	
Number and Operations-Fractions	**5.NF**
■ Use the equivalent fractions as a strategy to add and subtract fractions. ■ Apply and extend previous understandings of multiplication and division to multiply and divide fractions.	
Measurement and Data	**5.MD**
● Convert like measurement units within a given measurement system. ● Represent and interpret data. ■ Geometric measurement: understand concepts of volume and relate volume to multiplication and to addition.	
Geometry	**5.G**
◆ Graph points on the coordinate plane to solve real-world and mathematical problems. ◆ Classify two-dimensional figures into categories based on their properties.	

Figure 2.9 Sixth Grade Content Emphases

■—Major Cluster ●—Supporting Cluster ◆—Additional Cluster

Critical Focus Areas	
➡ Connecting ratio and rate to whole number multiplication and division and using concepts of ratio and rate to solve problems. ➡ Completing understanding of division of fractions and extending the notion of number to the system of rational numbers, which includes negative numbers. ➡ Writing, interpreting, and using expressions and equations. ➡ Developing understanding of statistical thinking.	
Expected Fluency	
➡ Divide multi-digit numbers and perform decimal operations on multi-digit numbers. (6.NS.2, 3)	
Domains and Clusters	
Ratios and Proportional Reasoning	**6.RP**
■ Understand ratio concepts and use ratio reasoning to solve problems.	
The Number System	**6.NS**
■ Apply and extend previous understandings of multiplication and division to divide fractions by fractions. ◆ Compute fluently with multi-digit numbers and find common factors and multiples. ■ Apply and extend previous understandings of numbers to the system of rational numbers.	
Expressions and Equations	**6.EE**
■ Apply and extend previous understandings of arithmetic to algebraic expressions. ■ Reason about and solve one-variable equations and inequalities. ■ Represent and analyze quantitative relationships between dependent and independent variables.	
Geometry	**6.G**
● Solve real-world and mathematical problems involving area, surface area, and volume.	
Statistics and Probability	**6.SP**
◆ Develop understanding of statistical variability. ◆ Summarize and describe distributions.	

Figure 2.10 Seventh Grade Content Emphases

■—Major Cluster ●—Supporting Cluster ◆—Additional Cluster

Critical Focus Areas

- ➡ Developing understanding of and applying proportional relationships.
- ➡ Developing understanding of operations with rational numbers and working with expressions and linear equations.
- ➡ Solving problems involving scale drawings and informal geometric constructions and working with two- and three-dimensional shapes to solve problems involving area, surface area, and volume.
- ➡ Drawing inferences about populations based on samples.

Expected Fluency

- ➡ Solve $px + q = r$ and $p(x + q) = r$.

Domains and Clusters

Ratios and Proportional Reasoning — **7.RP**

- ■ Analyze proportional relationships and use them to solve real-world and mathematical problems.

The Number System — **7.NS**

- ■ Apply and extend previous understandings of operations with fractions to add, subtract, multiply, and divide rational numbers.

Expressions and Equations — **7.EE**

- ■ Use properties of operations to generate equivalent expressions.
- ■ Solve real-life and mathematical problems using numerical and algebraic expressions and equations.

Geometry — **7.G**

- ◆ Draw, construct, and describe geometrical figures and describe the relationships between them.
- ◆ Solve real-life and mathematical problems involving angle measure, area, surface area, and volume.

Statistics and Probability — **7.SP**

- ● Use random sampling to draw inferences about a population.
- ◆ Draw informal comparative inferences about two populations.
- ● Investigate chance processes and develop, use, and evaluate probability.

Figure 2.11 Eighth Grade Content Emphases

■—Major Cluster ●—Supporting Cluster ◆—Additional Cluster

Critical Focus Areas	
➡ Formulating and reasoning about expressions and equations, including modeling an association in bivariate data with a linear equation and solving linear equations and systems of linear equations.	
➡ Grasping the concept of a function and using functions to describe quantitative relationships.	
➡ Analyzing two- and three-dimensional space and figures, using distance, angle, similarity, and congruence, and understanding and applying the Pythagorean Theorem.	
Expected Fluency	
➡ Solve simple 2 x 2 systems by inspection.	
Domains and Clusters	
The Number System	**8.NS**
● Know that there are numbers that are not rational and approximate them by rational numbers.	
Expressions and Equations	**8.EE**
■ Work with radicals and integer exponents.	
■ Understand the connections between proportional relationships, lines, and linear equations.	
■ Analyze and solve linear equations and pairs of simultaneous linear equations.	
Functions	**8.F**
■ Define, evaluate, and compare functions.	
■ Use functions to model relationships between quantities.	
Geometry	**8.G**
■ Understand congruence and similarity using physical models, transparencies, or geometry software.	
■ Understand and apply the Pythagorean Theorem.	
◆ Solve real-world and mathematical problems involving volume of cylinders, cones and spheres.	
Statistics and Probability	**8.SP**
● Investigate patterns of association in bivariate data.	

The Curriculum Focus Audit Tool

How aligned is your current curriculum to the instructional shift of focus? This curriculum focus-auditing tool can be used to determine the degree of alignment with content that should be emphasized at each grade level. It is important to evaluate the degree of alignment to determine appropriate time allocations in your current curriculum.

Curriculum Focus Audit Tool Protocol

Your current curriculum guide (scope and sequence) can be used to calculate the number of days allocated for each content cluster. Percentages can then be calculated to assess focus alignment.

Step 1: Make a copy of the Curriculum Focus Audit Tool template for your grade level. Blank templates are provided on the Digital Resource CD.

Step 2: Assuming your school runs on a 180-day calendar, determine the number of instructional days you may have available. (You should take into account testing days, school celebrations, etc.) This becomes your total number of instructional days (e.g., total = 180 – miscellaneous days).

Step 3: Use your curriculum and note the number of days you plan to spend (or days allocated by a district scope and sequence guide). The total overall percentage of time spent on the major clusters should fall in the 65% to 75% range. Exact percentages vary based on student need.

Step 4: If your time allocations are not aligned to the instructional shift of focus, you should redistribute allocated days to balance appropriately. (Balancing major clusters may require moving day allotment from supporting and additional clusters to major clusters and vice versa, or from major cluster to major cluster if there isn't equitable distribution of time between major clusters.)

Use the example shown in Figure 2.12 to help you get started.

Figure 2.12 Sample Curriculum Focus Audit Tool—Kindergarten

Major Cluster	Instructional Days Allocated	% of Time Allocated (# of Days ÷ Your Total # of instructional days) × 100
K.CC.A Know number names and the count sequence.	*30 days (2 standards)*	*30/165 x 100 = 18%*
K.CC.B Count to tell the number of objects.	*16 days (4 standards)*	*16/165 x 100 = 10%*
K.CC.C Compare numbers.	*18 days (2 standards)*	*18/165 x 100 = 11%*
K.OA.A Understand addition as putting together and adding to, and understand subtraction as taking apart and taking from.	*25 days (5 standards)*	*25/165 x 100 = 15%*
K.NBT.A Work with numbers 11–19 to gain foundations for place value.	*12 days (1 standard)*	*12/165 x 100 = 7%*

Supporting (s) and Additional (a) Clusters	Instructional Days Allocated	% of Time Allocated (# Days ÷ Your Total # of instructional days) × 100
K.MD.A Describe and compare measurable attributes. (a)	*7 days (2 standards)*	*7/165 x 100 = 4%*
K.MD.B Classify objects and count the number of objects in categories. (s)	*4 days (1 standard)*	*4/165 x 100 = 2%*
K.G.A Identify and describe shapes. (a)	*17 days (3 standards)*	*17/165 x 100 = 10%*
K.G.B Analyze, compare, create, and compose shapes (s)	*6 days (3 standards)*	*6/165 x 100 = 4%*
To be aligned to the Common Core, teachers should devote 65% to 75% of class time to content in the major clusters, with grades K–2 nearer the upper end of that range.		Does the time you allocated for the major clusters meet this criteria? Yes / (No)

Notes and Next Steps:

We have approximately 165 instructional days.

Only 61% of the time is planned for major clusters!

- *Need 9% or 15 additional days*

6% of the total is planned for supporting clusters.

- *Need 14% or 23 additional days*

14% of the time is planned for additional clusters.

- *Reduce by 4% or 7 days*

I notice that the total percentage of time is only 81%. 19% of the time is unaccounted for. That's approximately 31 days (unaccounted for) to add time to the supporting clusters and major clusters. I need to also consider the balance of time within the major clusters. Implications—add time to major clusters as needed based on how students are performing and supporting clusters.

I need to reduce the amount of time dedicated to the additional clusters. 17 days for the geometry unit is too much. I'll reduce this by 9 days. That will put me closer to the 10% goal for additional clusters.

Now I have a total of 9 days (taken from the additional clusters) and 31 additional days. 70-20-10!

You Try It!

1. Download the Curriculum Focus Audit Tool template for your grade level from the Digital Resource CD.
2. Review the Curriculum Focus Audit Tool Protocol (page 60).
3. Follow the protocol to complete the chart.
4. Determine whether your current grade-level curriculum is aligned to the instructional shift of focus. Discuss with your grade-level peers how to redistribute the number of instructional days spent on the major, supporting, and additional clusters.

The standards have been designed to focus on content that is key to meaningful learning at each grade level. Often curricular resources may highlight content based on publisher preference, responding more to popular demand and less to the content expertise described in the Common Core State Standards. The curriculum audit tool allows you to make informed decisions about time spent on required topics to ensure that students are prepared to negotiate mathematics at the next grade level. When the audit tool is complete, you are able to outline an instructional calendar that dedicates instructional time to the proper clusters as needed.

Stop and Think

Question 1

Why is it necessary to prioritize standards? How might you prioritize the standards at your grade level?

Question 2

How can you determine the degree of alignment with content that should be emphasized at your grade level?

Question 3

How does understanding focus impact your existing ideas regarding content taught at certain grade levels?

Chapter 3

Instructional Shift of Coherence

Mathematics is a discipline of interrelated concepts. These concepts are often taught in isolation, which reduces the chance that students understand the interdependence of topics. The instructional shift of coherence creates an opportunity for students to see the relationship between concepts, not only within a particular grade level, but also across grade levels. Cognitive learning theory suggests that meaningful learning occurs when new concepts are appropriately integrated into an existing schema (Gredler 2004). Integration occurs when connections are made between new concepts and existing knowledge structures. Students are then able to retrieve this information from their long-term memory. This chapter highlights practical tools that you can use to increase your understanding of the relatedness between mathematical concepts and to create an environment that allows students to construct these connections, too. In order to implement the instructional shift of coherence, you must have an understanding of the progression of mathematics. It will require you to understand the connections between concepts within your grade level as well as across grade levels. The following tools can help you make this shift in your classroom.

- **Progressions Traffic Signal Deconstruction**—The standards are built on a series of progressions. These progressions outline the development of mathematical concepts and serve as the framework that connects grade-level domains leading to conceptual categories at the high school level. This deconstruction tool helps illustrate coherence across grade levels.
- **Cluster Analysis Tool**—This analysis protocol enables educators to increase their understanding of coherence within their grade level.
- **Focus and Coherence Analyzer**—The instructional shift of focus addressed in Chapter 2 outlines the major points of emphasis at a unique grade level. Supporting clusters are also highlighted and should be connected to a major cluster. This analyzer demonstrates how to construct those connections that will be targeted in curriculum development.
- **Focus and Coherence Planner**—This tool highlights the major, supporting, and additional grade-level clusters. It serves as an instructional tool to consider time allocation and instructional strategies to target coherence.
- **Question Strategies**—A series of questioning strategies is presented that allows students to analyze concepts presented in the classroom. Strategic and purposeful creation of questions also serves to increase coherence in mathematics.

- **Sample Lessons**—How do teachers explicitly incorporate coherence strategies within lesson design? This section includes grade level examples that target coherence development. Support for lesson creation is found in subsequent chapters.
- **Constructing Analytic Questions**—Strategic questions can be used to help students understand the connections between mathematical concepts. Analytic questions ask students to form relationships. These question types are presented in this section alongside factual and evaluative question types.
- **Universal Concepts Matrix**—Universal concepts exist that show the relatedness between concepts not only within mathematics but also across all disciplines. This universal concepts matrix focuses on interdisciplinary connections.
- **Student Coherency Tools**—This section includes student resources, such as a Coherence Ladder, Coherence Cube, Coherence Mosaic, and Math Chains. These tools of engagement increase student inquiry and allow them to construct connections between concepts.

Coherence within Grade-Level Clusters

Standards are the ingredients of clusters.

—Phil Daro, CCSS-M, primary author

Coherence emphasizes connections across grade level standards and within grade-level standards. The clusters serve as the overarching umbrella that connects individual standards. Clusters are situated within domains that progress from one grade level to the next. Why is it important to emphasize these connections? In order to learn mathematics in a meaningful way, students must be able to connect new content with their existing prior knowledge. Learning theory, specifically Information Processing Theory (Pressley and Harris 2006), suggests that when students make these connections the content becomes situated in a student's long-term memory. It then can be retrieved as needed. Understanding the relatedness between math topics eliminates the need for memory mathematics. This common practice is rote learning and only registers in a student's short-term memory. Coherence, in essence, is the glue that connects mathematical learning.

Coherence Across Grade Levels

The standards are built on a series of progressions. These progressions outline the development of mathematical concepts and serve as the framework that connects grade-level domains leading to conceptual categories at the high school level. A team of mathematicians and teachers led by William McCallum (2013) created these progressions.

The hierarchal nature of the discipline of mathematics is presented in these progressions. Ideas that start at a particular grade level grow in sophistication to the high school level.

For example, it is often said that algebra is the gatekeeper to higher-level mathematics. There is a misconception that this gate appears for the first time around the eighth or ninth grade. We can see by looking at the progressions that every teacher from kindergarten through middle school teaches algebra. In other words, the algebra gate begins on the first day of kindergarten.

Domains in the primary grades evolve as students move to the middle grades and high school. For example, let's take a look at the domain Operations and Algebraic Thinking in Figure 3.1. Students in grades kindergarten through fifth grade learn how objects and terms are related and begin to express these relationships using symbols. These symbols are often blank spaces, stars, or boxes (e.g., 2 + ___ = 5). Students are asked to find an unknown quantity. As students move to middle school, this unknown quantity begins to take the form of an abstract variable. The domain of Expressions and Equations in grades 6 through 8 increases a student's understanding of Operations and Algebraic Thinking in grades K through 5 by now considering abstract relationships between the operations previously studied and variables and constants.

All of this knowledge from grades K through 8 is called upon for students to begin to formally explore the course of Algebra I at the high school level. Students continue to explore relationships between terms; however, these relationships take on many forms beyond standard linear connections.

In order for students to develop a meaningful and enduring understanding of mathematics and progress through gates such as algebra, teachers must create learning experiences that show how topics within and across domains are linked.

The progressions also highlight the foundational knowledge needed at a previous grade level in order to access and understand the new knowledge at the current grade level. Teachers can use this outline of knowledge in order to highlight prerequisite skill deficits that may be contributing to a student's lack of understanding about grade-level content.

High school teachers are often heard saying that middle school teachers do not prepare students properly and a focus on major ideas is not apparent. In turn, middle school teachers often discuss how elementary school teachers do not help students truly grasp mathematics. The progressions outline a series of batons that must be passed from one grade level to the next. In order for each grade level to make its respective pass, it should have an understanding of not only its leg of the race, but the outline of the entire race. The product of teachers understanding this seamless transition will result in students at each grade level through the high school level having a coherent understanding of mathematics. These students will be college and career ready when they graduate from high school.

Figure 3.1 Common Core Domain Coherence

<table>
<tr><th>Kindergarten</th><th>1</th><th>2</th><th>3</th><th>4</th><th>5</th><th>6</th><th>7</th><th>8</th><th>HS</th></tr>
<tr><td>Counting and Cardinality</td><td colspan="8"></td><td rowspan="3">Number and Quantity</td></tr>
<tr><td colspan="6">Number and Operations in Base Ten</td><td colspan="2">Ratios and Proportional Relationships</td><td></td></tr>
<tr><td colspan="3"></td><td colspan="3">Number and Operations—Fractions</td><td colspan="3">The Number System</td></tr>
<tr><td colspan="6" rowspan="2">Operations and Algebraic Thinking</td><td colspan="3">Expressions and Equations</td><td>Algebra</td></tr>
<tr><td colspan="2"></td><td>Functions</td><td>Functions</td></tr>
<tr><td colspan="6">Geometry</td><td colspan="3">Geometry</td><td>Geometry</td></tr>
<tr><td colspan="6">Measurement and Data</td><td colspan="3">Statistics and Probability</td><td>Statistics and Probability</td></tr>
</table>

(Ohio Department of Education 2012)

Examining the Progression of Skill and Knowledge Development

It is important for teachers to understand their grade-level expectations in order to spend the necessary time developing this foundational grade-level knowledge. A progression chart shows how skills advance from one grade level to the next.

For example, let's consider how the understanding of place value progresses from first grade through third grade. Place value starts in first grade under the domain of Number and Operations in Base Ten. The standards expect that students will be able to add within 100 using one-digit and two-digit numbers and understand that ones are added with ones and tens with tens.

In second grade, the expectation increases. Students are now expected to fluently add and subtract within 100. The idea of fluency suggests precision in a timely manner. They are also now expected to add and subtract within 1,000. Students should have the ability to add up to four two-digit numbers at this grade level.

In third grade, the expectation continues to develop. Fluency is now expected for addition and subtraction within 1,000. Multiplication of one-digit whole numbers by multiples of ten is also introduced.

One domain is unique to only a particular grade level: Counting and Cardinality in kindergarten. You can see from the chart in Figure 3.1 that this domain does not progress explicitly. The knowledge and skills acquired in this domain impact all of math, as counting serves as a fundamental building block. Conversely, one domain, Geometry, extends across all grade levels. Geometry means measurement of the earth. This includes a study of the world we live in by examining sizes, shapes, and relationships. We study mathematics in order to make sense of the world in which we live.

Another important reason why teachers should familiarize themselves with the progressions is to understand where new content is introduced on the spectrum of learning. For example, fractions begin in third grade with an understanding of where a fraction exists on a number line and the partitioning of a whole (1) into parts (b) producing $\frac{1}{b}$. In fourth grade, students begin to compare fractions with different numerators and denominators. And in fifth grade, multiplication and division of fractions is presented alongside mixed numbers.

The progressions charts found on the Digital Resource CD outline how the standards, clusters, and domains cross over multiple grade levels. An excerpt from the Progression Charts is shown in Figure 3.2.

Figure 3.2 Progressions Chart Excerpt

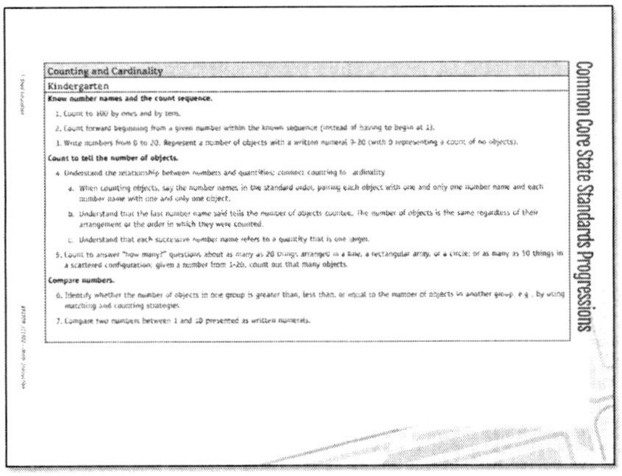

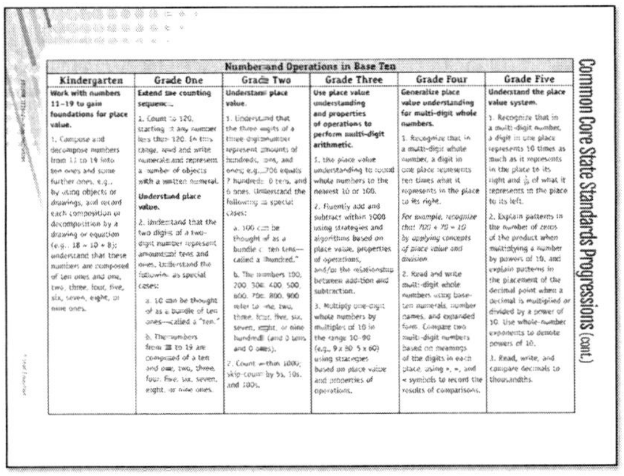

The following activity can help you better understand the progression of standards and in turn can create meaningful opportunities for your students to understand coherence. This can be done individually or in grade-level teams. For example, teachers in the same grade level choose a domain to analyze. Each teacher will then answer the questions in step 4 below. The grade-level team will share their findings and collaborate to reach a consensus. Their collective findings can be placed on a poster, and then each grade-level team can present their work.

You Try It!

1. Download the Progressions Charts from the Digital Resource CD (filename: progressions.pdf).
2. Start with a domain in your grade level and identify the key skills and knowledge targeted. Highlight these in yellow.
3. Identify the skills and knowledge needed in previous grade levels within the same domain that serve as prerequisites for your grade-level content. Highlight these in green.
4. Identify the skills and knowledge in future grade levels within the same domain. Highlight these in pink.
5. Respond to these questions:
    - How do the grade level expectations increase in complexity from one grade to the next? Describe the progression of the clusters in the selected domain.
    - How can these progressions be used to provide intervention for identified math deficits? List your ideas.

Analyzing Coherence within Clusters

It is important to examine the curriculum for coherence in order to be able to plan lessons that show connections between individual standards and the overarching cluster. The Cluster Analysis Tool in Figure 3.3 emphasizes the relationships not only between a standard and a cluster topic but also between the standards. Purposeful planning provides an opportunity to create lessons that explicitly incorporate these connections.

This tool can be completed individually or in grade-level teams. For example, each teacher on a grade-level team may choose one cluster and complete the analysis tool for that cluster. The grade-level team can then reconvene to discuss their findings. Each teacher then presents his or her ideas along with supporting reasons and evidence.

In order to create authentic lessons in which students understand the concepts and are not merely memorizing prescriptive procedures, teachers must spend time understanding how the standards are coherent. This tool can be used prior to creating a lesson to develop your own personal mathematics understanding. As a result, the opportunity for students to understand the connections increases.

Figure 3.3 Sample Cluster Analysis Tool

Domain: *Grade 4 Operations & Algebraic Thinking*

Cluster: *Use the four operations with whole numbers to solve problems.*

Standard 1: *4.OA.1*

How is this standard related to the cluster?

This standard highlights interpreting multiplicative equations used to compare.

How is this standard related to other standards in this cluster?

This standard emphasizes multiplication. 4.OA.2 incorporates the inverse operation—division. The equations are now used to solve word problems in 4.OA.2.

What will you include in your lesson to show these connections?

The cluster will be posted on the board. The lesson deals with multiplication to show comparisons. I will discuss with the class the other possible operations (e.g, cluster—four operations to solve).

Standard 2: *4.OA.2*

How is this standard related to the cluster?

Division is introduced. This is another one of the four operations mentioned in the cluster.

How is this standard related to other standards in this cluster?

We are now multiplying or dividing to solve word problems. Multiplicative comparisons are still being highlighted.

What will you include in your lesson to show these connections?

We will begin by interpreting a product such as $21 = 7 \times 3$ *as a statement that 21 is 7 times as many as 3. (4.OA.1) I will introduce an unknown symbol such as* n *and division.*

Example: *What is the relationship between the following sentence and the multiplicative comparison written above?*

$21 \div n = 3$

Standard 3: *4.OA.3*

How is this standard related to the cluster?

These problems include multi-step operations to solve with whole numbers. This is almost identical to the cluster.

How is this standard related to other standards in this cluster?

This standard uses both division and multiplication as referenced in 4.OA.1 and 4.OA.2. Word problems are embedded in 4.OA.2.

What will you include in your lesson to show these connections?

As we are solving the multi-step word problems, I will emphasize the multiplicative comparisons that are mentioned in 4.OA.1. Students will be asked to explain how the product compares with each factor in the word problem. If the product involves a quotient, students will be asked to explain the relationship between the quotient and divisor or dividend. Students will have to explain what the remainder represents in the context of the problem.

Chapter 3

You Try It!

1. Download the Cluster Analysis Tool template from the Digital Resource CD (filename: clustertool.pdf).

2. Select a domain to explore. Isolate a particular cluster within this domain.

3. Choose a standard and record how this standard relates to the cluster.

4. Record how this standard relates to other standards in the cluster.

5. Instructional Tool—What will you incorporate in your lesson to create an opportunity for students to understand these connections?

Focus and Coherence

The instructional shift of focus outlined in the previous chapter defines the major points of emphasis at a unique grade level. Coherence emphasizes that all math content is related. Supporting clusters may be connected to a major cluster in order to build coherence and help students see the connections among mathematical concepts. Creating an opportunity for students to make these connections requires purposeful planning to incorporate coherence.

The Cluster Emphasis chart highlights the combination of coherence and focus across grade levels. An excerpt of the chart is shown in Figure 3.4. A completed chart can be located on the Digital Resource CD.

Figure 3.4 Cluster Emphasis Chart

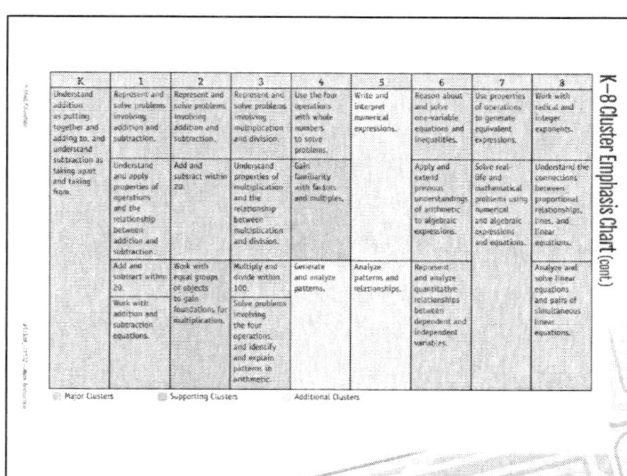

K–8 Cluster Emphasis Chart (cont.)

K	1	2	3	4	5	6	7	8
Understand addition as putting together and adding to, and understand subtraction as taking apart and taking from.	Represent and solve problems involving addition and subtraction.	Represent and solve problems involving addition and subtraction.	Represent and solve problems involving multiplication and division.	Use the four operations with whole numbers to solve problems.	Write and interpret numerical expressions.	Reason about and solve one-variable equations and inequalities.	Use properties of operations to generate equivalent expressions.	Work with radical and integer exponents.
	Understand and apply properties of operations and the relationship between addition and subtraction.	Add and subtract within 20.	Understand properties of multiplication and the relationship between multiplication and division.	Gain familiarity with factors and multiples.		Apply and extend previous understandings of arithmetic to algebraic expressions.	Solve real-life and mathematical problems using numerical and algebraic expressions and equations.	Understand the connections between proportional relationships, lines, and linear equations.
	Add and subtract within 20.	Work with equal groups of objects to gain foundations for multiplication.	Multiply and divide within 100.	Generate and analyze patterns.	Analyze patterns and relationships.	Represent and analyze quantitative relationships between dependent and independent variables.		Analyze and solve linear equations and pairs of simultaneous linear equations.
	Work with addition and subtraction equations.		Solve problems involving the four operations, and identify and explain patterns in arithmetic.					

Major Clusters   Supporting Clusters   Additional Clusters

Analyzing Focus and Coherence

This analyzer provides an opportunity for you to investigate the relatedness between clusters. It allows you to analyze how a supporting or additional cluster may connect to one or more major clusters in order to support lesson design. Lessons aligned to supporting clusters should reinforce the concepts addressed in major clusters. Additional clusters may not connect tightly to the major grade-level clusters, but instruction around additional clusters is still important.

Analysis of selected major, supporting, and additional clusters for each grade level are provided in the samples that follow. The rationale and application for each connection is provided.

Figure 3.5 Focus and Coherence Analyzer—Kindergarten

●—Supporting Cluster ◆—Additional Cluster

Measurement and Data	K.MD
◆ Describe and compare measurable attributes.	
● Classify objects and count the number of objects in categories.	
Geometry	**K.G**
◆ Identify and describe shapes.	
● Analyze, compare, create, and compose shapes.	

Supporting Cluster: Classify objects and count the number of objects in categories.	
Major Cluster: Compare numbers.	**Major Cluster:** Count to tell the number of objects.
Rationale: Counting the number of objects in a given group lends itself to comparing and ordering groups based on the number of elements.	**Rationale:** Counting the number of objects in each category involves counting to tell the number of objects present in each set.
Application: Have students count 20 items from a bag that can be classified in different ways (size, colors, shape, etc.). Instruct students to separate the items according to their selected classification. Then, have them count the number of items in each group. Allow students to repeat this multiple times using different classifications. (K.MD.3) Within this task, ask students to compare the quantities in each group. (K.CC.6)	**Application:** Prepare several triangles, squares, and circles of varying sizes and colors (limit to 3 different sizes and colors). Instruct students to sort the shapes first based on color, then based on size, and finally by shapes. Ask students to identify the number of objects in each group and to compare groups before they sort in a different way. (K.MD.3) Have students count the numbers out loud and make the connection that the last number represents the total number of objects. (K.CC.4a–b)

<table>
<tr><th colspan="2">Supporting Cluster: Analyze, compare, create, and compose shapes.</th></tr>
<tr><td>Major Cluster: Work with numbers 11–19 to gain foundations for place value.</td><td>Major Cluster: Count to tell the number of objects.</td></tr>
<tr><td>Rationale: The major cluster calls for decomposition of numbers between 11–19 using objects to see this relationship. Analyzing shapes requires counting terms. This can be used as a vehicle to see the decomposition.</td><td>Rationale: Composing shapes presents an opportunity to analyze the relationship between old and new shapes based on the number of sides and vertices. Counting to determine this and comparing the numbers is incorporated.</td></tr>
<tr><td>Application: Use several 3D shapes from home, including items such as cans of different sizes, shoe boxes, baseballs, footballs, basketballs, soda cans, cones, and triangular prisms. Have individual students pick any two objects and determine similarities and differences. Encourage students to use terms such as vertices and sides. Then, have students choose several objects with a common characteristic (e.g. eight vertices) and have students determine the rule or pattern. (K.G.4)

Make sure you have more than 10 items (11–19). Instruct students to count the total number of elements and decompose the total into a group of ten and additional ones. (K.NBT.1)</td><td>Application: Distribute two triangles of the same size, two rectangles of the same size, and a hexagon (size matches six of the triangle sizes included). Direct students to count the number of sides that each shape has. Direct students to put two triangles together to form a new shape. Count the number of sides of the new shape. Direct students to compose a new shape from placing two rectangles together, and a rectangle and a triangle. Have students think about decomposing the hexagon into triangles. How many triangles would be the result? (K.G.6)

As students are counting the sides, discuss the fact that each count represents a number that is one larger than the previous number. (K.CC.4c)</td></tr>
</table>

Figure 3.6 Focus and Coherence Analyzer—First Grade

●—Supporting Cluster ◆—Additional Cluster

Measurement and Data	1.MD
◆ Tell and write time. ● Represent and interpret data.	
Geometry	**1.G**
◆ Reason with shapes and their attributes.	

Supporting Cluster: Represent and interpret data.	
Major Cluster: Work with addition and subtraction equations.	**Major Cluster:** Add and subtract within 20.
Rationale: Interpreting data involves understanding the relationship between the different categories of data. The operations of addition and subtraction can be used to interpret this relationship numerically.	**Rationale:** Representing and interpreting data sets that involve sums that do not exceed 20 provides an opportunity to extend a student's comprehension of the operations of addition and subtraction.
Application: Choose a topic of interest to survey your students that has three different options from which to choose (e.g., favorite ice cream flavors, favorite sports). Ask students which of the three is their favorite. Record this data on the board in a chart. Ask the following questions: 1. How many total answers are present? 2. Which category has the most? How many more does this category have over the second place category? 3. What is the difference between the second and third categories? 4. How many more students like the most favored category over the least favored category? Have students explain their answers. (1.MD.4)	**Application:** Create a three-column chart on the board labeled *morning*, *afternoon*, and *evening activities*. Ask students to think about what activities they perform during these time periods. You may choose to determine more specific timeframes, if desired. Record student responses on the chart as they are presented. Ask students the following questions: 1. What time of day do we perform the most activities? 2. What time of day do we perform the least amount of activities? 3. How many more activities do we perform in ____ than in ____? 4. ____ category has how many fewer activities than ____? 5. What does this tell us about our day? Instruct students to explain their answers as they respond. (1.MD.4)

Chapter 3

Supporting Cluster: Represent and interpret data. (cont.)	
While students are responding to the questions, instruct them to write mathematical sentences that represent the question and the desired answer using addition and subtraction. For example, in question 3, after students have determined to subtract the values in each category, they may record the following sentence and interpret in the context of the problem. (e.g. 8 – 3 = ?) (1.OA.8)	Direct students to create sentences involving sums (e.g., 4 + 9 = 13) and corresponding differences (e.g., 13 – 9 = 4) Direct students to count on making ten when determining totals (e.g., 7 + 5 = 7 + 3 + 2 = 10 + 2) **Note:** Limit number of responses recorded so that potential addends will not produce a sum greater than 20. (1.OA.6)

Additional Cluster: Reason with shapes and their attributes.	
Major Cluster: Add and subtract with 20.	**Major Cluster:** Measure lengths indirectly and by iterating length units.
Rationale: Additional clusters are not tightly connected to the major work of the grade. In this case, the numbers of quarters and halves can be added to determine total parts in the whole. (e.g. 1 quarter + 3 quarters = 4 parts representing the whole, or 1 half + 1 half = 2 parts representing the whole)	**Rationale:** Additional clusters are not tightly connected to the major work of the grade. The tangram produces three different triangles of varying sizes. The lengths of the triangle sides can be compared using the length of a third object, the square in this case.
Application: Provide students with an activity sheet with two circles and two rectangles. Instruct students to draw lines that show one circle partitioned in halves and the second circle partitioned into quarters. Instruct students to partition one rectangle into halves and the other into quarters. How many different ways can a circle or rectangle be partitioned into quarters? Instruct students to describe the relationship between the parts and the whole using terms such as *quarter of* or *half of*. (1.G.3) Have students count the number of parts in the shape to find the total. (1.OA.5)	**Application:** Tangram puzzles can be used to help students understand how two-dimensional shapes can be used to create composite shapes. 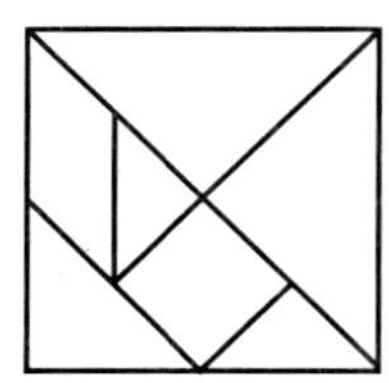 Cut tangram pieces and have students arrange the pieces to create a square, a triangle, and other classic tangram images. (1.G.2) Instruct students to compare the lengths of the three different triangles in the tangram puzzle by using the square. (1.MD.1)

Figure 3.7 Focus and Coherence Analyzer—Second Grade

●—Supporting Cluster ◆—Additional Cluster

Operations and Algebraic Thinking	**2.OA**
● Work with equal groups of objects to gain foundations for multiplication.	
Measurement and Data	**2.MD**
● Work with time and money. ● Represent and interpret data.	
Geometry	**2.G**
◆ Reason with shapes and their attributes.	

Supporting Cluster: Work with equal groups of objects to gain foundations for multiplication.	
Major Cluster: Represent and solve problems involving addition and subtraction.	**Major Cluster:** Relate addition and subtraction to length.
Rationale: Students are asked to count by 2s to determine whether a value is even or odd. Addition can be used to determine the sum of these groups of two. Students are asked to form equations using unknown addends to determine the even parts of an even number.	**Rationale:** Rectangular arrays are used to determine total sums. The arrays can be represented as rectangles that have a corresponding length and width. Total sums can be calculated using the noted length of each array.
Application: Instruct students to answer the following questions: 1. Are there an even or odd number of boys in the room? Are there an even or odd number of girls in the room? 2. Determine the number of students who like a particular sports team. Is that an even or an odd number? 3. How do we know a number is even? Instruct students to use the count by 2s rule. (e.g., if you can count all objects using 2s then the value is even). 4. Write an equation for all of the even quantities found using the sum of two unknown equal addends. What relationship do you notice about the addends and the resulting sum? (2.OA.3) Instruct students to complete the following equation frame related to the even numbers found in the previous activity: _____ + _____ = even number (2.OA.1)	**Application:** Have students form three different rectangular arrays of stars that include a total of 20 stars. Ask the following: What do these rectangular arrays have in common? How are they different? Have students explain how they determined the rectangular arrays, and write an equation that represents the total as a sum of equal addends. (2.OA.4) Explain to students that each star represents one unit of equally spaced length. Ask students to determine the dimensions of each rectangular array? Have them write each length as a sum of equal parts in an equation to show the total sum. (2.MD.6)

<table>
<tr><th colspan="2">Supporting Cluster: Work with time and money.</th></tr>
<tr><td>Major Cluster: Use place value understanding and properties of operations to add and subtract.</td><td>Major Cluster: Understand place value.</td></tr>
<tr><td>Rationale: When dealing with money, place value can be used to find the sum. For example, equating dimes with place value of tens allows for comprehension of not only addition with place values and multiples of ten but also adding money.
</td><td>Rationale: Skip counting by 5s is used to determine the time on the clocks. 2.NBT.5 calls for the use of skip counting by 5s to understand place value.</td></tr>
<tr><td>Application: Present students with the following problem scenario:
Elaine reached into her pocket and pulled out the coins that you see above. How much money does she have? Explain your thinking.
She went to the store and wanted to purchase a pack of gum that costs 50 cents and an eraser that costs 45 cents. Does she have enough money? Explain how you know. (2.MD.8)
Elaine collects dimes. If she did not spend her money in the store, what is the maximum number of dimes for which she can exchange her coins? (2.NBT.5)</td><td>Application: Present students with the following problem scenario:
Sandra works in a clock store. The owner has asked her to determine the times on the clocks so that they can be adjusted. What are the times on the clocks to the nearest 5 minutes?

Explain how you determined the time. (2.MD.7 and 2.NBT. 2)</td></tr>
</table>

Figure 3.8 Focus and Coherence Analyzer—Third Grade

●—Supporting Cluster ◆—Additional Cluster

Number and Operations in Base Ten	3.NBT
◆ Use place value understanding and properties of operations to perform multi-digit arithmetic.	
Measurement and Data	**3.MD**
● Represent and interpret data.	
◆ Geometric measurement: recognize perimeter as an attribute of plane figures and distinguish between linear and area measures.	
Geometry	**3.G**
● Reason with shapes and their attributes.	

Supporting Cluster: Represent and interpret data.	
Major Cluster: Develop an understanding of fractions as numbers.	**Major Cluster:** Solve problems involving the four operations, and identify and explain patterns in arithmetic.
Rationale: The standard being discussed involves halves and quarters on a ruler. These measurements represent numbers, namely fractions, that can be located on a number line and can be interpreted as the sum of unit fractions starting from 0.	**Rationale:** Interpreting data involves understanding how the different sets are related. Relationships may involve noticing patterns about the quantity, size, or value of the data sets. This provides an opportunity to incorporate the four operations to determine relatedness.
Application: Provide each student with a few commonly known items in a bag (paper clip, crayon, eraser, pen, etc.) Distribute rulers with halves and quarters of an inch marked on the ruler. Instruct students to measure the items in the bag and to record this information on a line plot. Instruct students to find other objects in the room (including trading objects from the bag with other students) and measure to the nearest half or quarter inch and record the data on their existing line plots. Direct students to only include items that fall in a designated range (e.g., under 6 inches). Which lengths are most represented? What items are these? What lengths are least represented? (3.MD.4)	**Application:** Create an activity sheet with varying rectangular bracelets (make sure that there are several bracelets that are over 4 inches in length for the second part of the problem) that measure different quarter and half inches. Instruct students to color and design each bracelet. Instruct students to measure each bracelet and record the lengths on a line plot. What bracelet is the longest length? How many bracelets share this length? How many more bracelets share the longest length than bracelets that share the shortest length? What additional questions do you have about the data collected? (3.MD.4) Present students with the following problem scenario:

Supporting Cluster: Represent and interpret data. (cont.)	
Choose three measured items whose lengths are greater than one. Plot these lengths on different number lines, marking off lengths of either $\frac{1}{2}$ or $\frac{1}{4}$ or the number (depending on the selected value—if the value was $\frac{5}{4}$ then the student would mark $\frac{1}{4}$s). What does the endpoint represent? (3.NF.2b)	The students at Linwood Elementary School plan on selling the bracelets that are over 4 inches. They want to determine the overall profit. Each 4-inch bracelet costs $2.00 to make. They can sell each bracelet over 4 inches for a value of $5.00. What will the total profit be if all bracelets over 4 inches are sold? Explain how you determined this result. (3.OA.8)

Supporting Cluster: Reason with shapes and their attributes.	
Major Cluster: Understand concepts of area and relate area to multiplication and addition.	**Major Cluster**: Develop an understanding of fractions as numbers.
Rationale: Determining attributes of shapes often involve understanding the number of sides present in a shape, vertices, lengths, and areas of shapes. Total area can be counted using square units and also found by multiplying features of the shape.	**Rationale:** Determining equal partitions of a particular shape and representing these partitions as fractions provides an opportunity for students to continue to develop their understanding of the relationship between fractions and parts of a whole.
Application: Provide students with an activity sheet with many shapes (circles, squares, rectangles, rhombuses, pentagons, etc.). Instruct students to classify the given shapes into two distinctive categories—quadrilaterals and non-quadrilaterals. Have them explain their reasoning. Have students identify additional quadrilaterals that are not present. What do all the quadrilaterals have in common? What differences exist among the quadrilaterals? (3.G.1) Distribute graph paper with square units. Instruct students to draw all of the quadrilaterals on the graph paper. Instruct the students to label the quadrilaterals with the appropriate names. How many square units make up the area of each quadrilateral? Instruct the students to count the square units and explain what patterns they notice. (3.MD.5, 6)	**Application:** 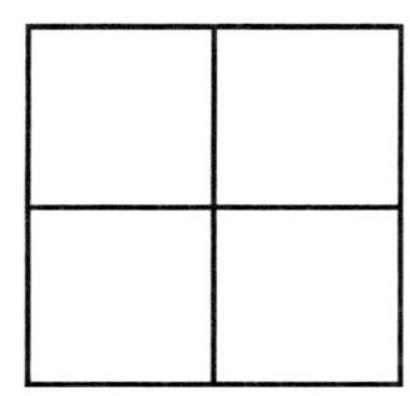 Present students with the following problem scenario: Susan wants to divide each square into 4 equal parts. How many different ways can she divide a square into 4 equal parts? Instruct the students to include more squares if they need more than 4 and explain their reasoning. (3.G.2) Choose one square that has been partitioned into four equal parts. How might you represent one of the parts as a fraction of the whole? How might you represent three of the parts as a fraction of the whole? How is the fraction determined from representing one part of the whole related to the fraction used to represent 3 parts of the whole? (3.NF.1)

Figure 3.9 Focus and Coherence Analyzer—Fourth Grade

●—Supporting Cluster ◆—Additional Cluster

Operations and Algebraic Thinking	4.OA
● Gain familiarity with factors and multiples. ◆ Generate and analyze patterns.	
Measurement and Data	**4.MD**
● Solve problems involving measurement and conversion of measurements from a larger unit to a smaller unit. ● Represent and interpret data. ◆ Geometric measurement: understand concepts of angle and measure angles.	
Geometry	**4.G**
◆ Draw and identify lines and angles, and classify shapes by properties of their lines and angles.	

Supporting Cluster: Gain familiarity with factors and multiples.	
Major Cluster: Use the four operations with whole numbers to solve problems.	**Major Cluster:** Generalize place value understanding for multi-digit whole numbers.
Rationale: When dealing with factors and multiples, students can continue to develop their understanding of the relationship between factors of a number and the number itself or multiples of a number and the number itself using the operations of multiplication and division.	**Rationale:** Understanding composite numbers provides an opportunity to compare multi-digit numbers using place value understanding. Inequality symbols and equality symbols can be used to express the comparison.
Application: Present students with the following problem scenario: Lauryn and Reagan sold Valentine's Day cookies. Lauryn made oatmeal raisin cookies and sold them in packs of three. Reagan made chocolate chip cookies and sold them in packs of 4. Both Lauryn and Reagan sold 48 cookies. How many packs of cookies did they each sell? Explain your reasoning. On a different day, they sold additional cookies packaged in the same manner. On this day, they both sold over 52 cookies. What are three different amounts that Lauryn may have sold? What are three different amounts that Reagan may have sold? Explain your reasoning. (4.OA.4) For the three quantities found for each girl, represent the relationship between the number sold (over 52) and the number of packs sold using equations with the unknown value. (4.OA.2)	**Application:** Present students with the following problem scenario: The Brown family was celebrating grandma Brown's 72nd birthday. It was determined grandma's age was a multiple of everyone at the party. What are the possible ages of all the people at the party? Determine if each age is a composite or prime number. Explain your reasoning. (4.OA.4) Have students compare the ages of all the adults at the party (adult being 18 or over) using inequality symbols or an equal symbol. Have them use the meaning of the digits in each place value in their explanation of the comparison. (4.NBT.2)

Supporting Cluster: Represent and interpret data.	
Major Cluster: Extend understanding of fraction equivalence and ordering.	**Major Cluster**: Build fractions from unit fractions.
Rationale: Interpreting data at this grade level involves comparing fractions with different denominators. This provides an opportunity to reinforce the concepts of finding common denominators or using benchmark fractions to compare fractions with different denominators.	**Rationale:** Representing and interpreting data involving the unit fractions of $\frac{1}{8}$, $\frac{1}{4}$,and $\frac{1}{2}$ provided an opportunity for students to continue to develop their understanding of simple fractions and how they are composed of unit fractions.
Application: Distribute rulers and instruct students to measure the length of their thumb using $\frac{1}{8}$s, $\frac{1}{2}$s, or $\frac{1}{4}$s. Record this data on the board. Instruct students to construct a line plot of the data. What do you notice about the data? What is the difference between the longest thumb and the shortest thumb (use equivalent fractions)? Explain how you solved this problem. (4.MD.4) Choose 4 pairs of sizes with different denominators and compare these results using inequality or equality symbols. (4.NF.2)	**Application:** Plot the following data on a line plot. $\frac{1}{4}$, $\frac{1}{2}$, $\frac{1}{4}$, $\frac{1}{2}$, 0, 1, $\frac{3}{4}$, $\frac{1}{4}$, $\frac{1}{2}$, 0, $\frac{1}{4}$, $\frac{3}{4}$, 0, 1, $\frac{3}{4}$, 0, $\frac{1}{4}$ What data point is represented the most? How is this reflected in the line plot? What value is represented the least? What is the difference in the number of times these two values are represented? What is the difference in the value of these terms? (4.MD.4) How is $\frac{3}{4}$ related to $\frac{1}{4}$? If these are measurements of lengths in inches, what is the total sum of all the $\frac{3}{4}$-inch measurements? Represent this total as a product in two different ways: A. Product involving the number of $\frac{3}{4}$ terms. B. Product involving the unit fraction $\frac{1}{4}$. Show that they are equivalent. (4.NF.4b)

Figure 3.10 Focus and Coherence Analyzer—Fifth Grade

●—Supporting Cluster ◆—Additional Cluster

Operations and Algebraic Thinking	5.OA
◆ Write and interpret numerical expressions. ◆ Analyze patterns and relationships.	
Measurement and Data	**5.MD**
● Convert like measurement units within a given measurement system. ● Represent and interpret data.	
Geometry	**5.G**
◆ Graph points on the coordinate plane to solve real-world and mathematical problems. ◆ Classify two-dimensional figures into categories based on their properties.	

Supporting Cluster: Convert like measurement units within a given measurement system.	
Major Cluster: Understand the place value system.	**Major Cluster:** Analyze patterns and relationships.
Rationale: Converting units of measurements in a given system requires understanding the relationship between numbers in certain place values.	**Rationale:** When students are converting like measurements, patterns are inherent that lead to generalizable rules.
Application: Present students with the following problem scenarios: A fence that runs along a portion of a farm has sections that are 4 meters long. If the portion of the farm is 0.4 km long, how many sections are there? Explain how you solved the problem. (5.MD.1) What patterns do you see while converting and determining the number of sections? How does this relate to place value? (5.NBT.2)	**Application:** Present students with the following problem scenarios: Ben has 7.3 L of punch to serve 40 people. How many milliliters of punch will each person receive? Explain how you solved the problem. (5.MD.1) Identify the relationship that exists between the amount of total punch, the number of people served, and the amount per person. Write a general rule that can be used to interpret this relationship. (5.OA.3)

Chapter 3

<table>
<tr><th colspan="2">Additional Cluster: Write and interpret numerical expressions.</th></tr>
<tr><td>Major Cluster: Perform operations with multi-digit whole numbers and with decimals to hundredths.</td><td>Major Cluster: Understand the place value system.</td></tr>
<tr><td>Rationale: Writing and interpreting numerical expressions can be learned by not only using single-digit whole numbers but also multi-digit whole numbers and decimals.</td><td>Rationale: Numbers written in expanded form serve as numerical expressions involving multiplication and addition that can be interpreted.</td></tr>
<tr><td>Application: Determine the value of the expressions:
a. (4,032 – 3,000) ÷ 400 × 5.5
b. 3.75 + 2.25 × 5
c. 12.5 × 40 + (60.24 ÷ 4.0)
(5.OA.1 and 5.NBT.5)</td><td>Application: Write numbers in expanded form and then instruct students to determine the value of the expression, which may result in a decimal.
Example: $4 \times 1{,}000 + 3 \times 100 + 8 \times 1 + 2 \times (\frac{1}{10}) + 5 \times (\frac{1}{100}) + 1 \times (\frac{1}{1000})$
(5.OA.1 and 5.NBT.3a)</td></tr>
</table>

Figure 3.11 Focus and Coherence Analyzer—Sixth Grade

●—Supporting Cluster ◆—Additional Cluster

The Number System	6.NS
◆ Compute fluently with multi-digit numbers and find common factors and multiples.	
Geometry	**6.G**
● Solve real-world and mathematical problems involving area, surface area, and volume.	
Statistics and Probability	**6.SP**
◆ Develop understanding of statistical variability. ◆ Summarize and describe distributions.	

Supporting Cluster: Solve real-world and mathematical problems involving area, surface area, and volume.	
Major Cluster: Apply and extend previous understandings of arithmetic to algebraic expressions.	**Major Cluster:** Reason about and solve one-variable equations and inequalities.
Rationale: Area, surface area, and volume can be found using procedural formulas. Determining comparable expressions that can be used to find area and volume may lead to students deriving commonly known formulas.	**Rationale:** When solving problems involving area, surface area, and volume, students can write equations that represent the relationships between the terms in the problem.
Application: Present students with the following problem scenario: You want to cover your trapezoid-shaped countertop using triangular-shaped slabs of marble. The countertop has a total area of 24 square feet. If 4 marble slabs are needed to cover the total area with no leftover, what are possible dimensions of one marble slab? Explain your solution. Include a drawing. (6.G.1) Write a numerical expression that can be used to determine the total area of the 4 slabs. Explain how this is useful in solving the problem. (6.EE.1)	**Application:** Present students with the following problem scenario: On a map represented by a coordinate plane, Powell Middle School is situated at (–6, 3), Dana Branch Library is located at (–6, –2), and Memorial Hospital is located at (4, –2). If each unit of measurement represents one mile, how far apart are Powell Middle School and Dana Branch Library? Firehouse 9 represents the fourth vertex that forms a rectangle with the other vertices. What are the coordinates of Firehouse 9? How far is Firehouse 9 from Memorial Hospital? Explain your reasoning. (6.G.3) Use a variable to represent the unknown lengths between the locations on the map and write an equation that can be used to determine the length. Interpret this equation in the context of the problem. (6.EE.6)

<table>
<tr><td colspan="2">Additional Cluster: Compute fluently with multi-digit numbers and find common factors and multiples.</td></tr>
<tr><td>Major Cluster: Apply and extend previous understandings of arithmetic to algebraic expressions.</td><td>Major Cluster: Apply and extend previous understandings of numbers to the system of rational numbers.</td></tr>
<tr><td>Rationale: Computing fluently with multi-digit numbers requires students to understand patterns that can be generalized. Developing an algebraic expression helps students to see these patterns.</td><td>Rationale: Multi-digit number computation may produce a rational number that is not whole. Finding an answer and situating this answer on a number line provides context for the solution.</td></tr>
<tr><td>Application: Present students with the following problem scenario:
Ms. Ashe's math class won the school mathematics challenge. The parents' association prepared 345 macaroons for the class. If 35 students in the class will receive an equal amount of macaroons, how many macaroons should each student receive? Explain your solution. (6.NS.2)
Write an algebraic expression that might be used to solve this problem. Write an equivalent version of the same expression. Explain why they are equivalent. (6.EE.4)</td><td>Application: Have students use a standard algorithm to perform the following operations:
20.32 + 17.97 = ____
17.45 × 6.9 = ____
72.6 ÷ 0.6 = ____
(6.NS.3)
Locate the resulting sum, product, or quotient on a number line. (6.NS.6)</td></tr>
</table>

Figure 3.12 Focus and Coherence Analyzer—Seventh Grade

●—Supporting Cluster ◆—Additional Cluster

Geometry	7.G
◆ Draw, construct, and describe geometrical figures and describe the relationships between them. ◆ Solve real-life and mathematical problems involving angle measure, area, surface area, and volume.	
Statistics and Probability	**7.SP**
● Use random sampling to draw inferences about a population. ◆ Draw informal comparative inferences about two populations. ● Investigate chance processes and develop, use, and evaluate probability models.	

Supporting Cluster: Use random sampling to draw inferences about a population.	
Major Cluster: Solve real-life and mathematical problems using numerical and algebraic expressions and equations.	**Major Cluster:** Analyze proportional relationships and use them to solve real-world and mathematical problems.
Rationale: Once random sampling is conducted to draw inferences about a population, equations can be constructed to produce data that supports the inference.	**Rationale:** When drawing inferences about a particular time period or range, this information can be used in conjunction with a proportion to determine a comparable range when conditions remain constant.
Application: Present students with the following problem scenario: Administration wants to increase the number of teachers who participate in the school's social events. The student council has been asked to conduct a survey of the teachers to determine their preferences for social events. The goal is to have at least 50% of the teachers attend social events. Which sampling method should the student council use? Why? Option 1: Pull random teacher's names from a box and survey those teachers. Option 2: Survey the first 10 teachers at the sign-in sheet. (7.SP.1) Write an equation including variables that can be used to determine the number of teachers needed to reach the goal. (7.EE.4)	**Application:** Present students with the following problem scenario: A bicycle store owner wants to track the number and types of bikes adults aged 21–25 purchase in his store each month. Which choice best represents a sample? A. Adults who buy bicycles B. 20 adults who are in the desired age range C. Adults in the desired age range who buy bikes D. Adults in the desired age range who come into the store (7.SP.1) Once the owner determines the number of bikes purchased in a month, he can establish a proportion that predicts the amount of a certain type of bike sold in a year, assuming the rate is constant every month. Write a proportion to represent this problem. (7.RP.2c)

<table>
<tr><td colspan="2">Additional Cluster: Solve real-life and mathematical problems involving angle measure, area, surface area, and volume.</td></tr>
<tr><td>Major Cluster: Use properties of operations to generate equivalent expressions.</td><td>Major Cluster: Solve real-life and mathematical problems using numerical and algebraic expressions and equations.</td></tr>
<tr><td>Rationale: Solving problems that involve area, circumference, and volume may require students to understand how to write equivalent expressions and how to rewrite commonly known formulas in terms of alternate variables in the formula.</td><td>Rationale: When solving for angle measurements, relationships between the known and unknown quantities may be established using equations and expressions. These equations are justified using angle properties and theorems.</td></tr>
<tr><td>Application: Display multiple circles and label the diameters or radii.
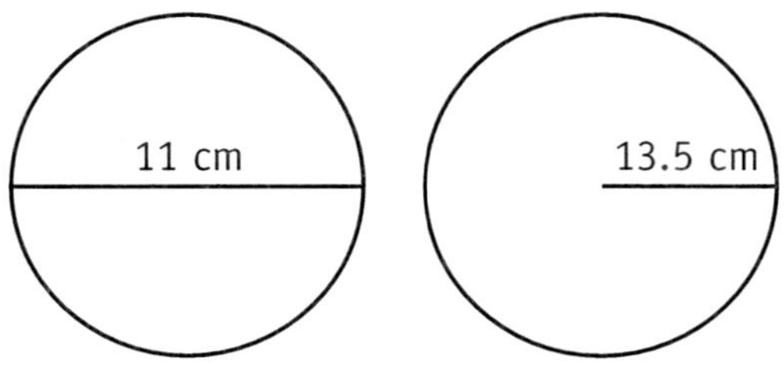
Instruct students to find the circumference and the area of the given circles. Then present them with the following questions:
How are the circumference and the area of a circle related? (7.G.4)
How might the circumference formula be written in two equivalent ways?
How might the area formula be expressed in terms of the circumference? (7.EE.1)</td><td>Application: Present students with the following problem:
Find the value of the unknown variables. Justify your solution using angle properties. (7.G.5)
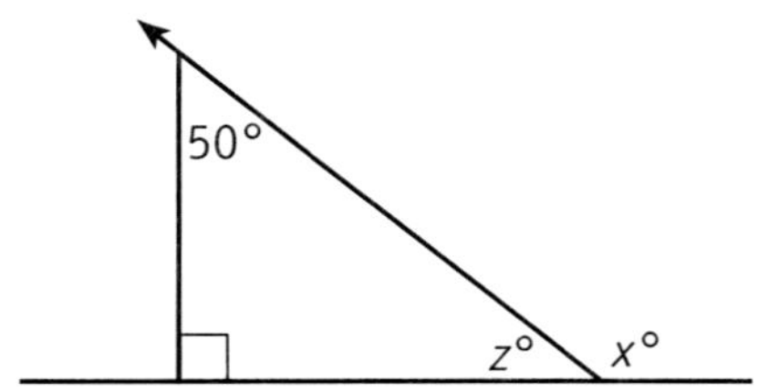
Construct at least three equations that show the relationship between the unknown angles and the known angles. Interpret the equations in the context of the problem. (7.EE.4)</td></tr>
</table>

Figure 3.13 Focus and Coherence Analyzer—Eighth Grade

●—Supporting Cluster ◆—Additional Cluster

The Number System	**8.NS**
● Know that there are numbers that are not rational, and approximate them by rational numbers.	
Geometry	**8.G**
◆ Solve real-world and mathematical problems involving volume of cylinders, cones, and spheres.	
Statistics and Probability	**8.SP**
● Investigate patterns of association in bivariate data.	

<table>
<tr><th colspan="2">Supporting Cluster: Know that there are numbers that are not rational, and approximate them by rational numbers.</th></tr>
<tr><td>Major Cluster: Work with radicals and integer exponents.</td><td>Major Cluster: Analyze and solve linear equations and pairs of simultaneous linear equations.</td></tr>
<tr><td>Rationale: Understanding the value of square roots that result in irrational numbers may require students to know how to evaluate perfect squares.</td><td>Rationale: There are equivalent fractions that represent rational repeating numbers. Systems of equations can be used to determine the equivalent fraction.</td></tr>
<tr><td>Application: Instruct students to draw a number line and place the following radicals on the number line:
$\sqrt{29}$, $\sqrt{2}$, $\sqrt{13}$
Have students explain their reasoning. (8.NS.2)
What perfect squares can be used to approximate the given radicals? Explain your selection. How might you approximate the value of the square root to the nearest tenth? (8.EE.2)</td><td>Application: Instruct students to represent a series of rational numbers as a fraction.
$0.3\overline{17}$
(8.NS.1)
Construct two linear equations that can be used to convert the rational number into a fraction. Explain your reasoning. (8.EE.8)</td></tr>
</table>

Supporting Cluster: Investigate patterns of association in bivariate data.	
Major Cluster: Use functions to model relationships between quantities.	**Major Cluster**: Understand the connections between proportional relationships, lines, and linear equations.
Rationale: Determining patterns that may exist between bivariate data presents an opportunity to construct formulas that can be used to predict a value of one quantity if the other quantity is known in a linear relationship.	**Rationale:** Investigating associations of bivariate data may reveal proportional relationships that can be used to determine best representative linear equations of the data.
Application: Have students investigate the following task: Is there an association between the length of a person's foot and their height? Have students measure their feet and height. Record this data on the board in a chart. Instruct students to form corresponding coordinates and plot the points on the coordinate plane. Are there outliers in the data? Does there appear to be a relationship between the two characteristics? Explain. Attempt to create a line that best represents this relationship. (8.SP.2) Construct a function that models the approximate linear relationship between the two quantities. (8.F.4)	**Application:** Present students with the following problem scenario: The growth rate of an oak tree is 2 in. per year after it was planted at a height of 20 feet. Write an equation that models this relationship. Graph the equation. What does the *y*-intercept represent? (8.SP.3) Describe the proportional relationship that exists. How might the unit rate of growth be used to graph the relationship? (8.EE.5)

You Try It!

1. Download the Focus and Coherence Analyzer for your grade level from the Digital Resource CD.
2. Review the sample in this chapter for your grade level.
3. Select different supporting/additional clusters to analyze and connect to the major clusters in your grade level. Or, use a supporting/additional cluster that has already been analyzed but connect it to a different major cluster.

Using the Focus and Coherence Planner

Now that you have spent time analyzing the connections among the major, supporting, and additional clusters at your grade level, it is important to examine your curriculum to see how the units are aligned to particular clusters and consider appropriate time allocations for each cluster. This tool provides an opportunity to consider the relatedness between standards in a cluster. In turn, this will increase your ability to create lessons that help students understand the relatedness between the standards. Instructional strategies that build coherence may also be noted and implemented in the corresponding unit of study. A sample of the Focus and Coherence Planner is provided in Figure 3.14.

Figure 3.14 Sample Focus and Coherence Planner—Grade 4

Common Core State Standards—Mathematics
Focus means that teachers will devote greater amounts of time to teaching the most important content in greater depth and with emphasis on the kind of complete understanding that will allow students to transfer their knowledge to other topics. ***Coherence*** means that teachers will devote time to making connections within and across grade-level content and linking to major topics within each grade. Standards are presented as extensions of previous information and not new isolated topics. In **Grade 4**, instructional time should focus on three critical areas: (1) developing understanding and fluency with multi-digit multiplication, and developing understanding of dividing to find quotients involving multi-digit dividends; (2) developing an understanding of fraction equivalence, addition and subtraction of fractions with like denominators, and multiplication of fractions by whole numbers; (3) understanding that geometric figures can be analyzed and classified based on their properties, such as having parallel sides, perpendicular sides, particular angle measures, and symmetry.

Explanations of Major, Additional, and Supporting Cluster-Level Emphases
Major clusters—areas of intensive focus where students need fluent understanding and application of the core concepts. These clusters require greater emphasis than the others based on the depth of the ideas, the time that they take to master, and/or their importance to future mathematics or the demands of college and career readiness. **Goal:** 70% of time; 126 total days **Actual:** ____ % of time ____ total days
Supporting clusters—rethinking and linking; areas where some material is being covered, but in a way that applies core understanding; designed to support and strengthen areas of major emphasis. **Goal:** 20% of time; 36 total days **Actual:** ____ % of time ____ total days
Additional clusters—expose students to other subjects; may not connect tightly or explicitly to the major work of the grade. *A Note of Caution: Neglecting material will leave gaps in students' skills and understanding and will leave students unprepared for the challenges of a later grade. **Goal:** 10% of time; 18 total days **Actual:** ____ % of time ____ total days

Operations and Algebraic Thinking	4.OA

Use the four operations with whole numbers to solve problems.

1. Interpret a multiplication equation as a comparison (e.g., interpret 35 = 5 × 7 as a statement that 35 is 5 times as many as 7 and 7 times as many as 5). Represent verbal statements of multiplicative comparisons as multiplication equations.
2. Multiply or divide to solve word problems involving multiplicative comparison (e.g., by using drawings and equations with a symbol for the unknown number to represent the problem, distinguishing multiplicative comparison from additive comparison).
3. Solve multistep word problems posed with whole numbers and having whole-number answers using the four operations, including problems in which remainders must be interpreted. Represent these problems using equation with a letter standing for the unknown quantity. Assess the reasonableness of answers using mental computation and estimation strategies, including rounding.

Unit/Section/Topic #	Instructional Days
2.3 Using Mental Math for Addition (4.OA.3)	*2 days*
3.1 Concept of Multiplication (4.OA.1, 4.OA.2)	*3 days*
3.2 Understanding Patterns (4.OA.1)	*1 day*
3.3 Properties of Multiplication (4.OA.1)	*1 day*
3.4 Factors of 3, 4, and 6 (4.OA.3)	*1 day*
3.6 Concept of Division (4.OA.2, 4.OA.3)	*3 days*
3.7 Multiplication/Division Relatedness (4.OA.1)	*2 days*
3.8 Unique Quotients (4.OA.2)	*1 day*
4.4 Problem Solving (4.OA.1, 4.OA.3)	*2 days*
4.6 Multistep Word Problems (4.OA.3)	*1 day*
7.2 Quotient Estimation (4.OA.3)	*1 day*
7.11 Multistep Problem Solving (4.OA.3)	*1 day*

Total Instructional Days: *19 days*

Coherence: How are the standards related to each other?

The standards use the four operations to solve problems involving whole numbers. Comparisons between values are made using multiplication and division.

How are they related to previous/future standards?

In third grade, students were asked to interpret products as the total number of a set number of groups and items in each group. Students used multiplication and division within 100 to solve word problems using arrays.

In fifth grade, students will be asked to use parentheses and brackets to evaluate expressions

Curricular/Instructional Strategies (e.g., Coherence Ladder, Coherence Mosaic, Math Chain, Questioning)

In lessons 3.1–3.8, we will use a coherence ladder to describe the concept being taught and discuss the connections between the previous concepts.

I will introduce the idea of order of operations that they will explore in the future. This note has been added to my textbook.

Operations and Algebraic Thinking	4.OA
Gain familiarity with factors and multiples.	
4. Find all factor pairs for a whole number in the range of 1–100. Recognize that a whole number is a multiple of each of its factors. Determine whether a given whole number in the range of 1–100 is a multiple of a given one-digit number. Determine whether a given whole number in the range of 1–100 is prime or composite.	
Unit/Section/Topic#	**Instructional Days**
7.9 Factors (4.OA.4) *7.10 Composite and Prime Numbers (4.OA.4)*	*2 days* *1 day*
	Total Instructional Days: *3 days*

Coherence: How are the standards related to each other?

There is only one standard in this cluster. However, finding factors relates to 4.OA.1—interpreting multiplication equations as a comparison. The connection can be made that 21 is 7 times as much as 3 relates two factors of 21 to show a comparison.

How are they related to previous/future standards?

3.OA.6 Understand division as an unknown-factor problem. For example, find 32 ÷ 8 by finding the number that makes 32 when multiplied by 8.

Extend upon this idea by understanding that the number found is a factor. The terms "prime" and "composite" numbers are introduced in 4.OA.4. Students can begin to understand the relationship between the number of whole number factors and the classification of a number as prime or composite.

Curricular/Instructional Strategies (i.e. Coherence Ladder, Coherence Mosaic, Math Chain, Questioning)

Will incorporate analytical questions throughout these two lessons.

What is the relatedness between composite and prime numbers and the number of factors a number may have?

What is the relationship between the quotient, dividend, divisor, and the concept of factors?

Building Coherence by Asking Sound Questions

Creating opportunities for students to understand the coherence in mathematics involves purposeful selection of instructional strategies. The intentional selection of strategies includes an understanding of explicit questions that can be incorporated into a learning experience in order to build coherence. Teachers must be aware of the implications of the questions they ask and the impact on a student's ability to determine the relatedness between the concepts being taught throughout a course of study. Using a range of question types provides space for a student to link old topics with new topics and understand overarching relationships in mathematics.

Questions can be categorized in many ways. One such way is open questions (divergent) versus closed questions (convergent). Within each category, there are more specific descriptions. All question types have a time and place in curriculum development and instructional delivery. There is no hard and set rule regarding when to use a particular question type. This is determined based on your understanding of balancing the frequency of question types used, participation of your students (open vs. closed access), desire for targeted specific information (factual), and timing (e.g., Would it be beneficial to ask a series of closed questions at the beginning of an introductory topic?).

To increase opportunities for coherence, analytical questions play an important role in highlighting relationships that exist. These questions compare math topics or procedures and make connections explicit. Factual, analytical, and evaluative question types follow. There are distinctive patterns for each question type. What patterns do you notice?

Figure 3.15 Types of Questions

Closed Questions		
Question Type	**Definition**	**Examples**
Factual	Questions that solicit verifiable facts as answers.	What is the measure of angle C? If you multiply 7 and 4, what is the resulting product? How many children are in the room? What is the solution of $x + 3 = -6$?
Open Questions		
Question Type	**Definition**	**Examples**
Analytical	Questions that target the relatedness between concepts examining part-to-whole and part-to-part relationships.	How does the measure of angle C relate to the area of the shape? How does the addition sentence 5 + 6 = 11 relate to the subtraction sentence 11 – 6 = 5? How are the different types of quadrilaterals similar and different? What do the terms of the quadratic equation indicate about the possible roots?
Evaluative	Questions that assess the value of a topic based on perspective.	How important is it to find the measure of angle C when finding the area of the shape? How do you determine what to do first when solving an inequality? Why are you using this drawing to represent the number of total apples? Why is it important to know the order of operations?

Constructing Analytic 21 Questions

As mentioned earlier, analytical questions increase opportunities for students to see the coherence within mathematics. Highlighting the relationship between topics builds a student's mathematical schema. The following question stems and descriptions may be used to construct analytical questions.

- How does the current topic build upon previous topics?
- How do the two topics ____________ and ____________ compare? How do they contrast?
- Explain the relatedness between the topics ____________ and ____________.

Constructing Evaluative Questions

Evaluative questions give students the opportunity to consider the importance and relevance of a math topic. The value of understanding a concept is often increased when students understand the relevance. The question stems and descriptions that follow may be used to construct evaluative questions:

- What is the importance of understanding the topic ___________ with the current topic of study?
- What is the relevance of the topic ___________?
- Why are you using your selected strategies to solve the problem? Why do you think it is a good approach?

1. **Consider an upcoming lesson/unit.**
2. **Prepare a list of analytical and evaluative questions that can be asked that increase coherency within the unit.**
3. **Consider the overarching cluster and the nuisances of the content when constructing your questions.**
4. **Tag these questions in your curricular resource.**

The following sample lessons incorporate coherence-building questions. In the grades K–2 sample lesson, students subtract multiples of 10 from multiples of 10 using concrete models or drawings. The grades 3–5 sample lesson focuses on solving problems involving measurement and conversion of measurements from a larger unit to a smaller unit. In the grades 6–8 sample lesson, students divide fractions by using visual fraction models and equations to represent the problem. Constructing Common Core inquiry-based math lessons will be addressed in more depth in Chapter 5.

Lesson Sample: *Subtracting Multiples of Ten*

Grades K–2

Cluster(s):

Use place value understanding and properties of operations to add and subtract.

Emphasis: **Major** (circled)
Supporting
Additional

Standards:

- Subtract multiples of 10 in the range 10–90 from multiples of 10 in the range 10–90 (positive or zero differences), using concrete models or drawings and strategies based on place value, properties of operations, and/or the relationship between addition and subtraction; relate the strategy to a written method and explain the reasoning used. (1.NBT.6)
- Make sense of problems and persevere in solving them. (MP1)
- Reason abstractly and quantitatively. (MP2)
- Model with mathematics. (MP4)

Learning Objective:

Students will be able to relate multiples of ten using concrete models or drawings.

Present the Problem/Cluster

1. Present the following mathematical scenario and cluster for students to investigate. Say, "We've been discussing the Common Core cluster of using place value understanding and properties of operations to add and subtract. Let's explore the following problem."
   Lauren has 40 pens. Ten of the pens have no ink. She gave ten of the pens with ink to her sister. How many pens does Lauren have that she can write with? Use a drawing to explain your reasoning.
2. Allow students time to discuss the intended meaning of the cluster and ask clarifying questions about the problem scenario.

Solicit Questions

3. Ask students to generate questions for exploration based on the task. Record questions that students generate.
4. If needed or desired, ask the following strategic questions to prompt exploration:
   - What is known in the problem? (MP1)
   - How does this problem relate to our cluster topic of place value? (Coherence)
   - How might you use a drawing to explain your thinking? (MP2)

Lesson Sample: *Subtracting Multiples of Ten* (cont.)

Grades K–2

Mathematical Research

5. Have students work in groups to develop strategies/methods to solve the problem and answer the questions that have been generated.
6. Circulate around the room while students are working and ask questions targeting the math practices, coherence (connection to cluster and other content), and rigor (concept—why, purpose; procedure—selected algorithm). Suggested questions:
   - Explain how your drawing represents the connection between all of the pens. (MP4)
   - How might you express the connection using both addition and subtraction? (Coherence)
   - How do you explain your answer in terms of the problem? (MP1)

Share and Summarize

7. Allow students to present their findings from research and their argument. Consider selecting students who illustrate specific solution pathways and solutions that are connected to the solicited questions from earlier in the lesson.
8. Allow students to act as respectful critics of each other's work by asking the following questions at various points throughout the presentations.
   - What do the different drawings have in common? (MP3)
   - Do you agree or disagree? Why? (MP3)

Recycle and Forecast

9. Highlight the studied cluster from the lesson and allow students to discuss the meaning that was created. Ask, "What is the relationship between adding and subtracting multiples of ten and what you know about place value?"
10. Ask students, "What additional questions remain for future exploration?" (Coherence, MP3)
11. Record any additional questions that students generate as a result of the research.

NOTES

Lesson Sample: *Comparing Units of Measurement*

Grades 3–5

Cluster(s):

Solve problems involving measurement and conversion of measurements from a larger unit to a smaller unit.

Emphasis: Major
(Supporting) [circled]
Additional

Standards:

- Use the four operations to solve word problems involving distances, intervals of time, liquid volumes, masses of objects, and money, including problems involving simple fractions or decimals, and problems that require expressing measurements given in a larger unit in terms of a smaller unit. Represent measurement quantities using diagrams such as number line diagrams that feature a measurement scale. (4.MD.2)
- Attend to precision. (MP6)
- Look for and make use of structure. (MP7)

Learning Objective:

Students will be able to compare distance measurements using a real-world scenario.

Present the Problem/Cluster

1. Present the following mathematical scenario and cluster for students to investigate. Say, "We've been discussing the Common Core cluster of solving word problems involving measurement and conversion of measurement from a larger unit to a smaller unit. Today's scenario gives us an opportunity to continue to understand this cluster topic."
   Michael ran $2\frac{1}{2}$ kilometers before he needed to stop and rest. Susie ran 5 laps around the track, and then she stopped. One lap is 500 meters. Who ran a longer distance? Justify your answer.
2. Allow students time to discuss the intended meaning of the cluster and ask clarifying questions about the problem scenario.

Solicit Questions

3. Ask students to generate questions for exploration based on the task. Record questions that students generate.
4. If needed or desired, ask the following strategic questions to prompt exploration:
   - What do you notice about the units of measurement in the problem? (MP6)
   - What do you know from the previous day's lesson that may be incorporated into this problem? (Coherence)
   - How does this problem relate to our overarching cluster? (Coherence)
   - How might you relate the different units of measurement? (MP7)

Lesson Sample: *Comparing Units of Measurement* (cont.)

Grades 3–5

Mathematical Research

5. Have students work in groups to develop strategies/methods to solve the problem and answer the questions that have been generated.
6. Circulate around the room while students are working and ask questions targeting the math practices, coherence (connection to cluster and other content), and rigor (concept—why, purpose; procedure—selected algorithm). Suggested questions:
   - How do you justify this result? (MP3)
   - Describe the procedure using proper academic language. (MP6)
   - Describe the relationship between the quantities represented in the problem. (Coherence)

Share and Summarize

7. Allow students to present their findings from research and their argument. Consider selecting students who illustrate specific solution pathways and solutions that are connected to the solicited questions from earlier in the lesson.
8. Allow students to act as respectful critics of each other's work by asking the following questions at various points throughout the presentations.
   - What do the solution pathways have in common? (MP3)
   - Do you agree or disagree? Why? (MP3)

Recycle and Forecast

9. Highlight the studied cluster from the lesson and allow students to discuss the meaning that was created. Ask, "What is the relationship between this concept and the previous concept of comparing units of measurement?" (Coherence)
10. Ask students, "What additional questions remain for future exploration?" (Coherence, MP3)
11. Record any additional questions that students generate as a result of the research.

NOTES

Lesson Sample: *Dividing Fractions*

Grades 6–8

Cluster(s):

Apply and extend previous understandings of multiplication and division to divide fractions by fractions.

Emphasis: (Major) [circled]
Supporting
Additional

Standards:

- Interpret and compute quotients of fractions, and solve word problems involving division of fractions by fractions, e.g., by using visual fraction models and equations to represent the problem. (6.NS.1)
- Make sense of problems and persevere in solving them. (MP1)
- Construct viable arguments and critique the reasoning of others. (MP3)
- Look for and make use of structure. (MP7)
- Look for and express regularity in repeated reasoning. (MP8)

Learning Objective:

Students will be able to apply division of fractions properties to solve and explain real-world problems.

Present the Problem/Cluster

1. Present the following mathematical scenario and cluster for students to investigate. Say, "Today we are continuing to explore our Common Core cluster to apply and extend previous understandings of multiplication and division to divide fractions by fractions."
   A jeweler has an $8\frac{1}{4}$ inch gold wire that she is cutting into pieces that are $1\frac{3}{8}$ inches. How many pieces can she cut? Explain how you determine this result.
2. Allow students time to discuss the intended meaning of the cluster and ask clarifying questions about the problem scenario.

Solicit Questions

3. Ask students to generate questions for exploration based on the task. Record questions that students generate.
4. If needed or desired, ask the following strategic questions to prompt exploration:
   - How would you describe the problem in your own words? (MP1)
   - What do you know from the previous day's lesson that may be incorporated into this problem? (Coherence)
   - How does this problem reinforce our cluster understanding of "apply and extend previous understanding of multiplication and division to divide fractions by fractions?" (Coherence)
   - What is known and unknown? (MP1)

Lesson Sample: *Dividing Fractions* (cont.)

Grades 6–8

Mathematical Research

5. Have students work in groups to develop strategies/methods to solve the problem and answer the questions that have been generated.
6. Circulate around the room while students are working and ask questions targeting the math practices, coherence (connection to cluster and other content), and rigor (concept—why, purpose; procedure—selected algorithm). Suggested questions:
   - How do you prove this result? (MP3)
   - Describe the relationship between these quantities? (MP7, Coherence)
   - What patterns do you see? (MP3, MP8)

Share and Summarize

7. Allow students to present their findings from research and their argument. Consider selecting students who illustrate specific solution pathways and solutions that are connected to the solicited questions from earlier in the lesson.
8. Allow students to act as respectful critics of each other's work by asking the following questions at various points throughout the presentations.
   - What do the solution pathways have in common? (MP3)
   - Do you agree or disagree? Why? (MP3)

Recycle and Forecast

9. Highlight the studied cluster from the lesson and allow students to discuss the meaning that was created. Ask, "What is the relationship between this concept and the previous concept of multiplication of fractions?" (Coherence)
10. Ask students, "What additional questions remain for future exploration?" (Coherence, MP3)
11. Record any additional questions that students generate as a result of the research.

NOTES

Universal Concepts Across the Curriculum

Universal concepts exist that show the relatedness between concepts not only within mathematics but also across all disciplines. Mathematics often exists on an academic island at school sites. Understanding coherence includes not only connections within and across grade-level mathematics but also across many disciplines.

Universal concepts are overarching ideas that connect learning across multiple disciplines, which ultimately increase coherence in mathematics. These universal concepts increase the complexity of content within an area of study (Kaplan and Curry 1985). The concepts allow students to understand the relatedness between mathematics and other subject areas. Universal concepts are not specific topics unique to mathematics. Individual facts are situated within large ideas known as concepts. Collections of concepts are situated in even larger ideas known as universal concepts. For example, a *point* is a fact that belongs to concepts such as polygons. *Polygons* are concepts that are defined by *structure*, which is a universal concept. The concept of *structure* can be defined across many disciplines, which makes it universal. Figure 3.16 is a descriptive list of universal concepts for which many subjects converge.

Figure 3.16 Universal Themes and Generalizations

Change (transform or convert)
- Generates additional change
- Can either be positive or negative
- Is inevitable
- Is necessary for growth
- Can be evolutionary or revolutionary

Conflict (contradict or variance)
- Is composed of opposing forces
- May be neutral or man-made
- May be intentional or unintentional
- May allow for synthesis and change
- Is progressive

Exploration (investigation)
- Requires recognizing purpose and responding to it
- Confronts the unknown
- May result in new findings or confirmation of old findings

Force (strength or intensity)
- Attracts, holds, or repels
- Influences or changes
- Force and inertia are co-dependent
- May be countered with equal or greater force

Order (methodical arrangement)
- May be natural or constructed
- May allow for prediction
- May have repeated patterns
- Order and chaos are reciprocals
- Order leads to chaos and chaos leads to order

Patterns (consistent configuration)
- Have segments that are repeated
- Allow for prediction
- Have an internal order
- Are enablers

Power (control or influence)
- Is the ability to influence
- May be used or abused
- Is always present in some form
- May take many forms

Structure (arrangement of parts or elements)
- Has parts that interrelate
- Has parts that support and are supported
- May combine to form larger structures
- Is no stronger than its weakest component

Systems (assembly of parts into a whole)
- Have parts that work together to complete a task
- Are composed of sub-systems
- Follow rules
- May be influenced by other systems
- Interact

Relationships (connection or association)
- Everything is related in some way
- Are powerful
- Are purposeful
- Change over time
- Follow rules

(Adapted from Kaplan and Curry 1985))

Incorporating Universal Concepts into the Curriculum

Universal concepts may be selected and used throughout the year (or term). Classroom or grade-level selection should take into consideration not only the mathematics content addressed throughout the year but also the content across all subject areas for that grade level. The selected universal concept and how it is represented in different disciplines should be reinforced periodically throughout the year (or term).

The following is a recommended sequence for incorporating universal concepts into the curriculum.

1. Discuss the selected universal concept with the class at the beginning of the year. Include the purpose of using universal concepts.
2. Provide generalizations about the universal concept and encourage students to create their own generalizations as it relates to the content (see Figure 3.15).
3. Create a living class poster that can be referenced throughout the year and updated as content is covered and new ideas are discovered.
4. Make connections (and encourage students to make connections) to the universal concepts throughout the year while covering units inside and outside of mathematics.

This practice is most beneficial when teachers across subject areas who share the same students reinforce the same universal concept as it relates to their subject throughout the term.

The sample lesson template on page 106 can be used to introduce a universal concept in the study of mathematics. The suggested syntax is the same for all grade levels (e.g., quick write/quick speak, solicit questions, student research, students answer questions, share and summarize, then recycle and forecast), although consideration should be given to the use of visuals and less complex language/vocabulary when working with young students.

Lesson Sample: *Universal Concepts*

Learning Objective:

Students will be able to determine the relevance of incorporating Universal Concepts into the study of mathematics. Students will use their math journals to record their findings.

Quick Write:

1. Present the following mathematical prompt: Use your journals to record the mathematics topics you studied last year. Include formulas, symbols, and graphic illustrations in your descriptions.
2. Allow students time to record their responses in their journals. Lower grade levels can discuss the prompt as a class, while the teacher records student responses on the board using words or drawings based on the grade level.

Solicit Questions

3. Ask students to generate questions for exploration based on the need to understand how the topics that have been written/spoken are connected. Record questions that students generate.
4. Ask the following strategic questions to prompt exploration:
    - *What connections are there between the topics you have listed?* (Students can choose two or more topics and discuss how they are related. This can be written or spoken based on grade-level readiness.)
    - *If these topics represent* ***concepts****, what might* ***universal concepts*** *be?*
5. After listening to the responses, the teacher then provides the following information:
    - A universal concept is a concept that encompasses multiple complex concepts made up of simple topics.
    - For example: simple concept/topic = truck, complex concept = transportation, universal concept = systems. Ask, "How is a truck related to the larger concept of transportation? How is transportation related to the larger universal concept of systems?" (Possible responses: A truck is a vehicle that might be used during transportation. Transportation is a type of organized representation of connected parts known as a system.)

Lesson Sample: *Universal Concepts* (cont.)

Mathematical Research

6. Have students work in groups to explain their understanding of the universal concepts. (Lower grade levels may choose to work as a class to share ideas about universal concepts.)
7. Present the universal concepts chart and post each concept on the board as it is introduced and discussed. (Lower grade levels should limit the number of universal concepts discussed. Perhaps only discuss the one or two universal concepts that will be used throughout the year. Teachers may want to find illustrations that represent each universal concept in addition to showing the word).
8. Give students an opportunity to explain what each universal concept may mean. Ask questions targeting the concepts. Suggested question:
   - *What universal concepts can be used to show the connections between the mathematics concepts you have written in your journals (or that are written on the board)?*
9. Present selected pictures that may represent various universal concepts, and ask students, "What universal concepts come to mind when you see these pictures? Explain."

Share and Summarize

10. Allow students to present their findings from research and their argument. Consider selecting students who illustrate specific solution pathways and solutions that are connected to the solicited questions from earlier in the lesson.
11. Present previous grade-level topics. Ask students to connect each topic to a universal concept.
12. List math topics that are familiar to students on the board, using illustrations. Examples may include a number line, fraction bars, operations, or a cartesian plane.

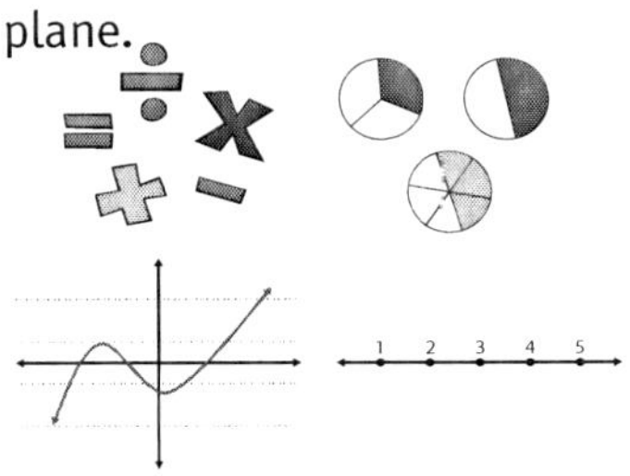

Recycle and Forecast

13. Highlight the targeted mathematics cluster from the current lesson and allow students to discuss the meaning of the cluster and how it relates to a selected universal concept.
14. Say, "This year we will use universal concepts to show how math is connected within the subject and across other subject areas you will study. This is called coherence. (Explain to lower grades what coherence means.) The selected universal concept will be posted on the board and we will reference it throughout the year."
15. "What additional questions might you have as a result of this lesson?" Record any additional questions that students generate as a result of the research.

Involving Students in the Coherence Shift

You cannot underscore the importance of students making connections between mathematical topics as mentioned in the introductory section of this chapter. The Standards for Mathematical Practice emphasize this point by including an understanding of structure and how objects and concepts are related as a measure of student proficiency.

Identifying Cross-Curricular Connections

The Universal Concept Matrix, shown in Figure 3.17, can be used as a planning tool to make connections during the delivery of instruction. Students can also complete the matrix to note connections between topics discussed. With a renewed emphasis on interdisciplinary connections and problem-based learning projects that span multiple disciplines, this universal concept matrix may be used as a guide throughout a unit of instruction. Students may use this tool throughout the unit to note connections from one subject to the next. This tool provides an opportunity for students to understand coherence not only within mathematics but also across other disciplines.

This chart can be used in multiple ways. The teacher can enlarge this chart and post it on the wall to serve as an organizing tool throughout the year. Students can incorporate this tool into their journal and add to it throughout the year.

In middle school, the mathematics teacher should consult with other subject-specific teachers to determine an appropriate universal concept(s) that aligns with the content scope and sequence for a particular semester/grading period.

Once this is determined, the teacher records the corresponding Common Core clusters as they are presented in the unit of study. Throughout the course of study, math concepts are recorded in the column on the right. Students are then encouraged to describe how these math concepts are connected to other subject areas discussed in other periods or segments throughout the day. The discipline row can be adjusted based on a student's particular course of study. The Universal Concept Matrix template can be found on the Digital Resource CD.

Figure 3.17 Universal Concept Matrix

Universal Concept					
Common Core Clusters					
	Connection to Other Disciplines				
Math Topic	**Science**	**English**	**History**	**Art**	**Other**

Constructing Knowledge One Rung at a Time

The Coherence Ladder provides an opportunity for students to construct connections between daily topics, Common Core clusters, and universal concepts.

Students can fill in each slot of the ladder as shown in Figure 3.17 to describe the topic being addressed. They can fill in the rungs of the ladder to describe the connections between the topics. The rails of the ladder can be used to list and describe the mathematical cluster.

The Coherence Ladder template can be found on the Digital Resource CD. A sample is provided in Figure 3.18.

Figure 3.18 Sample Coherence Ladder—Grade 5

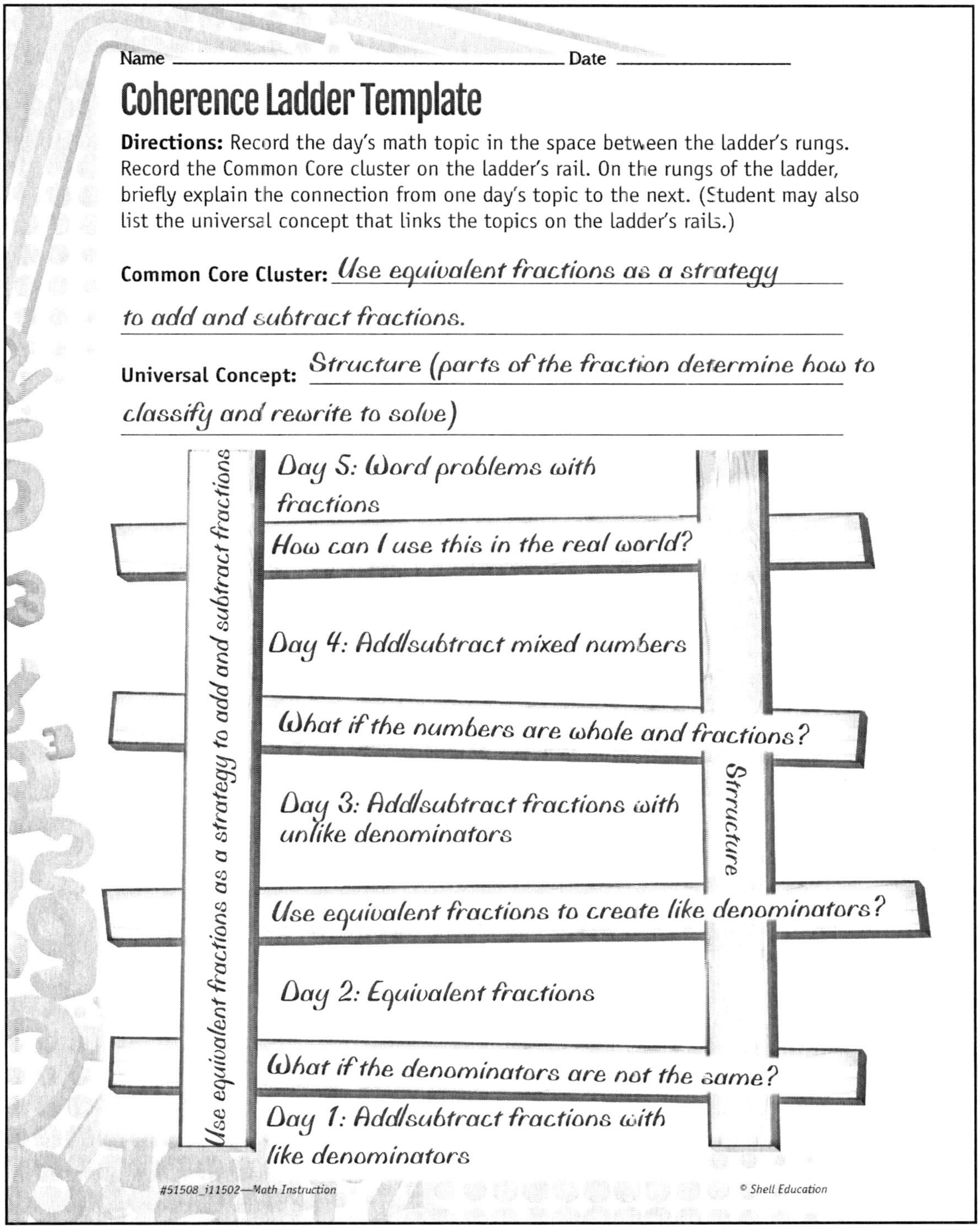

Name ______________________ Date ____________

Coherence Ladder Template

Directions: Record the day's math topic in the space between the ladder's rungs. Record the Common Core cluster on the ladder's rail. On the rungs of the ladder, briefly explain the connection from one day's topic to the next. (Student may also list the universal concept that links the topics on the ladder's rails.)

Common Core Cluster: *Use equivalent fractions as a strategy to add and subtract fractions.*

Universal Concept: *Structure (parts of the fraction determine how to classify and rewrite to solve)*

#51508_i11502—Math Instruction © Shell Education

Connecting the Present with the Past and Future

The Coherence Cube may be used to provide opportunities for students to discuss the relatedness between math concepts. Students roll the cube shown in Figure 3.19 and record how their current mathematics lesson is connected in the manner that appears. Students present their arguments in small groups. This can be used as assessment criteria.

The cube should be printed and constructed using tape. Students roll the cube and answer the corresponding questions, which may be posted on the board or provided in an accompanying handout. A template for the Coherence Cube can be found on the Digital Resource CD.

Figure 3.19 Sample Coherence Cube

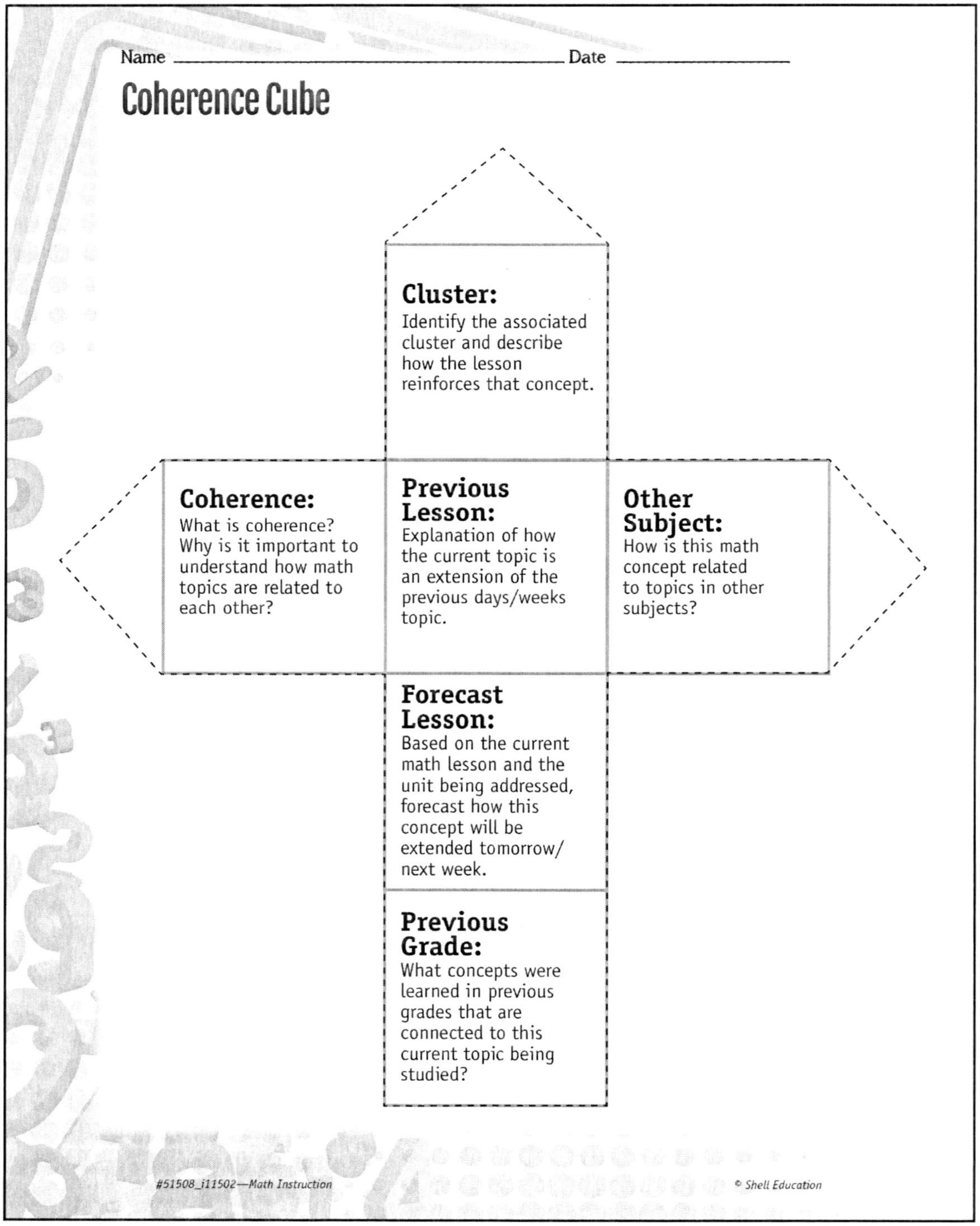

Name ______________________ Date ____________

Coherence Cube

Cluster:
Identify the associated cluster and describe how the lesson reinforces that concept.

Coherence:
What is coherence? Why is it important to understand how math topics are related to each other?

Previous Lesson:
Explanation of how the current topic is an extension of the previous days/weeks topic.

Other Subject:
How is this math concept related to topics in other subjects?

Forecast Lesson:
Based on the current math lesson and the unit being addressed, forecast how this concept will be extended tomorrow/next week.

Previous Grade:
What concepts were learned in previous grades that are connected to this current topic being studied?

#51508_i11502—Math Instruction *© Shell Education*

Seeing the Big Picture

Students can create a Coherence Mosaic over the course of the year or a semester. They start with a blank poster board and draw the mathematical concept being taught for a particular week or unit. On the back of the poster board, the student explains in words what the concept represents. Each new drawing is accompanied by an explanation on the back of how it is connected to the previous or other drawings on the mosaic as is evident on the grade 4 sample shown in Figure 3.20.

Figure 3.20 Sample Coherence Mosaic—Grade 4

Front:

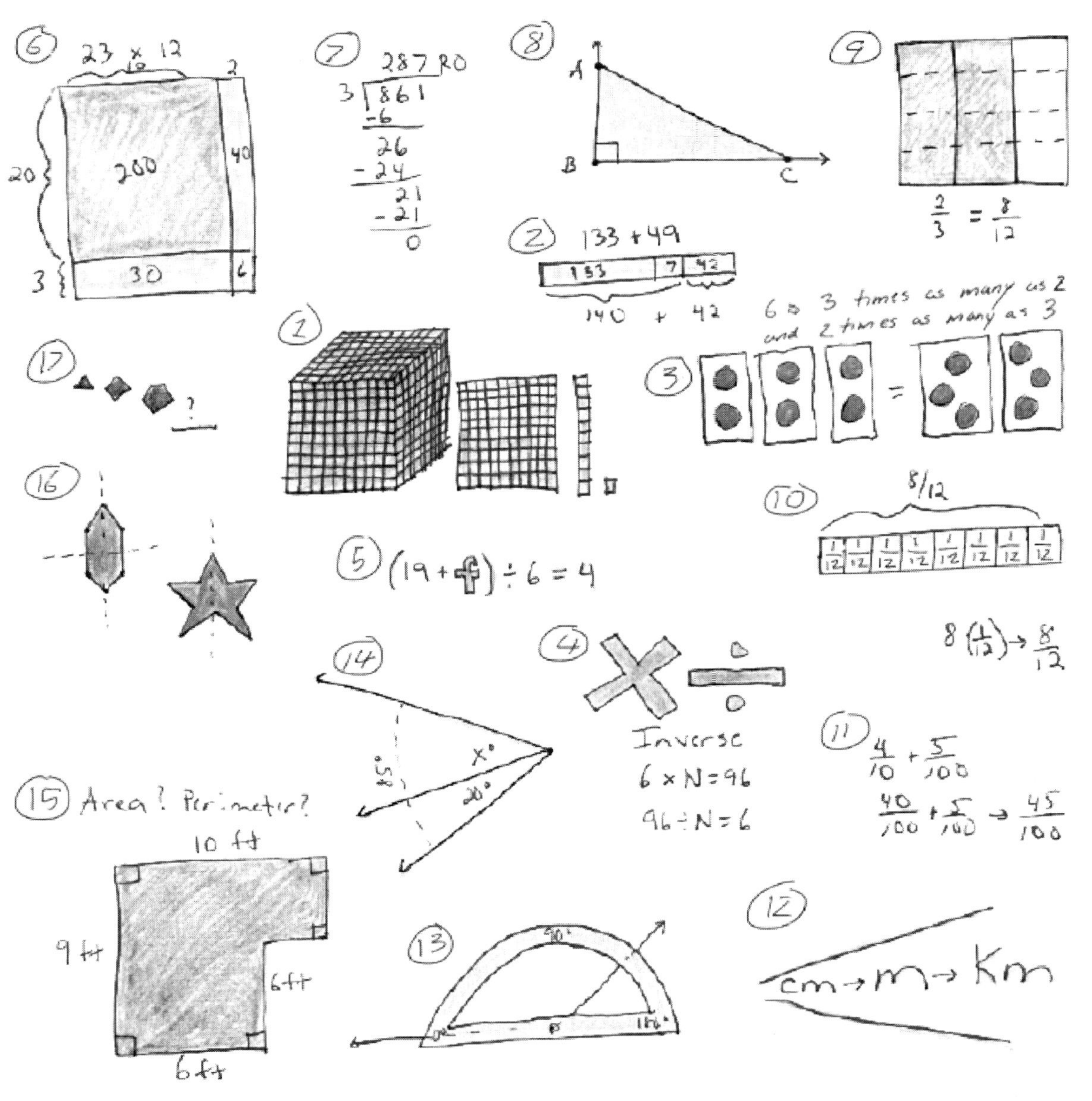

Back:

1. Place value understanding is used to compare numbers.
2. Use place values and sets of tens to help add.
3. Multiplying in any order results in the same answer. The answer is a comparison between the numbers multiplied.
4. Multiplication and division are inverse operations that undo each other.
5. Equations using operations and an unknown number can be used to solve problems.
6. Strategies to multiply numbers can be used, such as expanded form.
7. Finding quotients and explaining the remainder. How does this relate to factors of a number?
8. Right triangles have a 90-degree angle. The triangle is made up of points and segments. A segment may be extended to form a ray.
9. How are fractions with different denominators really equivalent? The whole is cut into different parts!
10. A simple fraction with a numerator other than one can be written as the sum of the same unit fractions or the unit fraction multiplied by the number of total unit fractions.
11. Find equivalent fractions of denominators using multiples of ten.
12. Attend to precision—units of measurement are related and increase/decrease in size.
13. Sketch angles and understand parts of an angle. Measure the angle in degrees using a protractor.
14. An angle can be split into smaller angles. The sum of these smaller angles equals the value of the larger angle.
15. Use the perimeter and area formula to understand real-world problems.
16. Symmetry can be found by identifying lines that can be folded so that the two halves overlap exactly. Shapes can have more than one line of symmetry.
17. Find rules that describe patterns. Use the rules to make predictions.

The Chain of Mathematics

Venn diagrams have often been used not only in mathematics but also across many disciplines to show how concepts compare and contrast. Creating opportunities for students at all grade levels to understand the similarities and differences between math concepts increases the likelihood that students will be able to see mathematics as a coherent subject. This also aids in retention of math concepts presented.

With a math chain, students record the math concepts on the activity sheet (see Figure 3.21). The intersecting region of the Venn diagram is used to record common features of the two math concepts being analyzed. Students complete this task independently, in small groups, or the teacher can record student responses (including writing for lower grade levels if necessary) in a whole-group setting. Certain grade levels may choose to use the rectangles at the bottom of the page as links of a Math Chain. To do this, the teacher or the student records representative ideas of each math concept. These links can be cut out and attached to create a chain. As new math topics are introduced throughout the year, the chain continues to grow. Students should discuss the connections between the links, using the Venn diagram.

A samp.e Math Chain is shown in Figure 3.21. The Math Chain template can be located on the Digital Resource CD.

Figure 3.21 Sample Math Chain—Grade 3

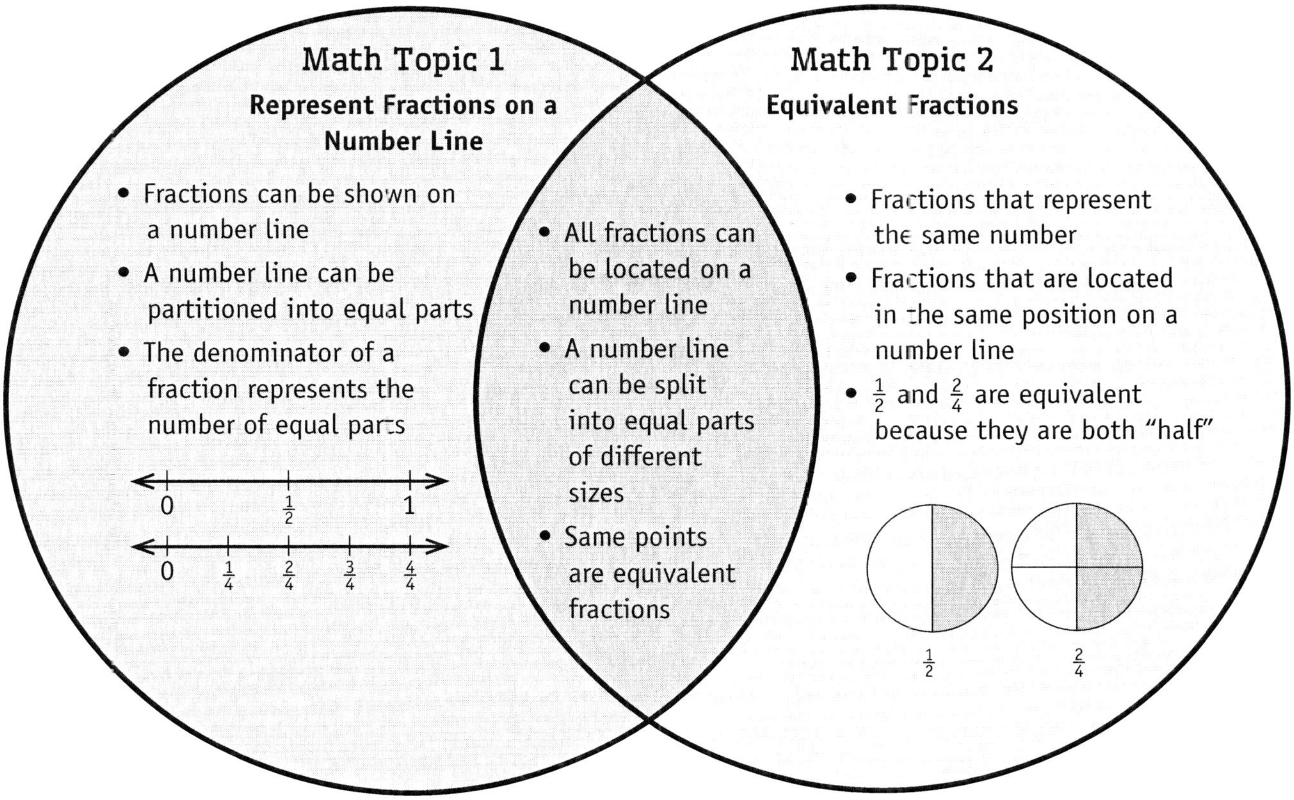

Math Topic 1: Represent Fractions on a Number Line

Description: Fractions are numbers on a number line. The number line can be split into equal parts that are equal to the denominator.

Math Topic 2: Equivalent Fractions

Description: Equivalent fractions share the same space on a number line.

Coherence Puzzle

The Coherence Puzzle gives students an opportunity to make connections between mathematical concepts throughout the school year. Here are a couple of different ways to use the Coherence Puzzle in your classroom:

Option 1: Teachers can record the mathematical concepts taught throughout the semester (or year) on the individual puzzle pieces. Students can then complete the puzzle by fitting pieces and discussing the connection between adjoining pieces.

Option 2: Students can use the puzzle mat throughout the semester (or year) to record concepts presented. At the end of the term, students can exchange puzzles for completion.

The Coherence Puzzle template can be found on the Digital Resource CD. Figure 3.22 shows a sample Coherence Puzzle for Grade 4.

Figure 3.22 Sample Coherence Puzzle—Grade 4

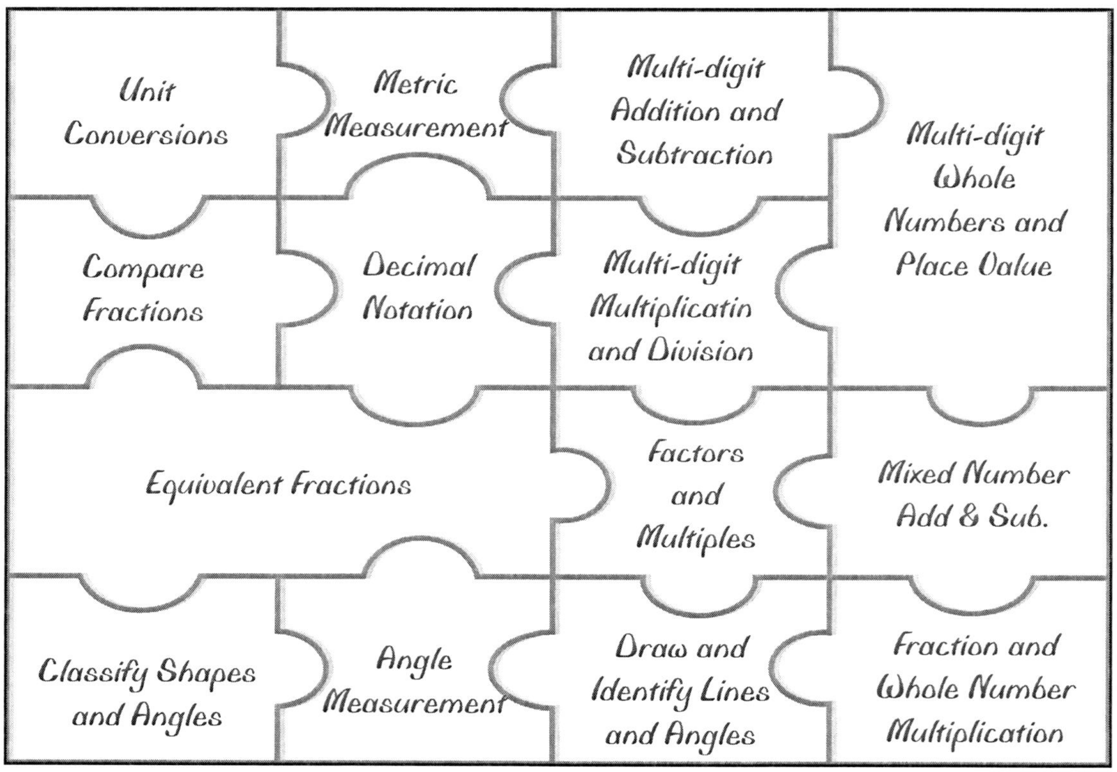

Stop and Think

Question 1

What strategies might you use to emphasize cluster knowledge in addition to the daily learning objectives of the lesson?

Question 2

How can you make the topic of coherence explict for students? Why is this important?

Question 3

Meaningful learning occurs when students make connections between content topics. How does this fact impact your motivation to include coherence strategies in your lesson design?

Chapter 4

Instructional Shift of Rigor

The term rigor is often heard in educational conversations relating to curriculum and instruction. When asked what it means, responses include challenging, difficult, higher-ordered thinking, and complex. However, tasks which may be challenging to one student may not be challenging to another suggesting that rigor is relative to the individual. Rigor is a key shift associated with the Common Core State Standards in Mathematics and has been clearly defined as the combination of the following dimensions: conceptual understanding, procedural skill, and application.

Figure 4.1 Dimensions of Rigor

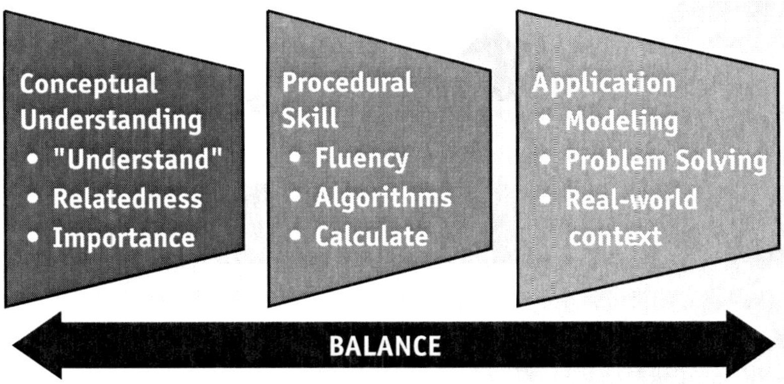

Conceptual Understanding: One dimension of rigor is conceptual understanding—the understanding of relationships between mathematical topics and concepts. Conceptual knowledge is an understanding of ideas, relationships, connections, or having a sense of something (Barr et. al. 2003). Conceptual knowledge targets purpose, importance, and value. The Common Core suggests that standards that emphasize the word "understand" are good starting points for conceptual exploration of a topic.

Procedural Skill: A second dimension of rigor is procedural skill and fluency. With the advent of the Common Core there is an increased emphasis on conceptual understanding. However, this emphasis is not at the exclusion of procedural knowledge. Rittle-Johnson and Alibali (1999) refer to procedural knowledge as action steps to derive answers. Procedural knowledge targets a fluent command of the operational functions. The word "fluent" is explicitly used in the standards. Use of algorithms to perform calculations and speed and accuracy in calculations are components of procedural knowledge.

Application: The third dimension of rigor is application, in which mathematical concepts are contextualized through modeling and problem solving. Mathematics is a language to make sense of the world in which we live. Procedural knowledge and conceptual knowledge should be applied in real-world contexts. Students should have an understanding of where this knowledge can be applied to negotiate a scenario. Mathematical Practice Standard 4, Model with Mathematics, provides a vehicle for application.

Star (2002) argues that instead of debating over superiority between conceptual and procedural knowledge, education needs to consider the relationships that exist between these two approaches to mathematics. Conceptual and procedural knowledge exist on a learning continuum and cannot be separated.

It is important to note as rigorous units of study are created that all three components must be treated with equal intensity. There is no particular sequence, nor does one dimension carry more weight than the other dimensions. All three elements may not be apparent in one lesson, but when exploring a particular topic, all three should be balanced.

This chapter includes the following.

- **Grade-Level Problem Types**—What does rigor look like in the context of math problems? Grade-level problems are provided that demonstrate each dimension and the characteristics are highlighted.
- **Rigor Calibration Tool**—This resource provides an opportunity for teachers to calibrate their understanding of various problem types. In turn, teachers will have the ability to determine what problems target a particular domain of rigor.
- **Rigor Analysis Protocol**—The standards directly call for different dimensions of rigor based on the language. This deconstruction tool provides an analysis of rigor in the standards.
- **Rigor Monitoring Tool**—Rigor should be explored across all content in areas in a given grade level with greater emphasis on the identified major clusters that target the grade level foci. The Rigor Monitoring Tool will help educators maintain a balanced program of rigor.

Identifying Rigor in the Standards

The standards are written to develop a student's comprehensive knowledge of mathematics. As previously stated, this includes conceptual knowledge, procedural knowledge, and application of this knowledge in a real-world context. There is no hard-and-steady rule that dictates what standards primarily emphasize a particular knowledge domain. However, the standards provide indicators that are helpful when determining points of emphasis. The word "understand" in a standard points towards a more in-depth conceptual understanding of the topic. The word "fluent" or the explicit inclusion of a mathematical operation (add, subtract, multiply, divide) suggests that an emphasis on procedural knowledge might be dominant. Standards that address real-world scenarios or modeling are good starting points to emphasize the application domain. The figures that follow include examples of standards that primarily target a particular knowledge domain. Keep in mind that targeting a particular knowledge domain does not mean that you completely exclude the other domains.

Figure 4.2 Examples of Knowledge Targets

Grade	Conceptual Knowledge Target Standard	Procedural Fluency Target Standard	Application Target Standard
K	**Understand** that each successive number name refers to a quantity that is one larger. (K.CC.4c partial standard)	Fluently add/subtract within 5. (K.OA.5)	Represent addition and subtraction with objects, fingers, mental images, drawings, sounds (e.g., claps), acting out situations, verbal explanations, expressions, or equations. (K.OA.1)
1	**Understand** the meaning of the equal sign, and determine if equations involving addition and subtraction are true or false. (1.OA.7)	Fluently add/subtract within 10. (1.OA.6)	Solve word problems that call for addition of three whole numbers whose sum is less than or equal to 20 (e.g., by using objects, drawings, and equations with a symbol for the unknown number to represent the problem). (1.OA.2)
2	**Understand** that in adding or subtracting three-digit numbers, one adds or subtracts hundreds and hundreds, tens and tens, ones and ones; and sometimes it is necessary to compose or decompose tens or hundreds. (2.NBT.7 partial standard)	Fluently add/subtract within 20. (2.OA.2) Fluently add/subtract within 100. (2.NBT.5)	Solve word problems involving dollar bills, quarters, dimes, nickels, and pennies, using $ and ¢ symbols appropriately. *Example: If you have 2 dimes and 3 pennies, how many cents do you have?* (2.MD.8)
3	**Understand** division as an unknown-factor problem. (3.OA.6)	Fluently multiply/divide within 100. (3.OA.7) Fluently add/subtract within 1,000. (3.NBT.2)	Solve real-world and mathematical problems involving perimeters of polygons, including finding the perimeter given the side lengths, finding an unknown side length, and exhibiting rectangles with the same perimeter and different areas or with the same area and different perimeters. (3.MD.8)

Grade	Conceptual Knowledge Target Standard	Procedural Fluency Target Standard	Application Target Standard
4	**Understand** addition and subtraction of fractions as joining and separating parts referring to the same whole. (4.NF.3a)	Fluently add/subtract within 1,000,000. (4.NBT.4)	Solve word problems involving multiplication of a fraction by a whole number (e.g., by using visual fraction models and equations to represent the problem). *For example, if each person at a party will eat $\frac{3}{8}$ of a pound of roast beef, and there will be 5 people at the party, how many pounds of roast beef will be needed? Between what two whole numbers does your answer lie?* (4.NF.4c)
5	**Understand** that attributes belonging to a category of two-dimensional figures also belong to all subcategories of that category. (5.G.3)	Fluently multiply multi-digit whole numbers. (5.NBT.5)	Solve real-world problems involving multiplication of fractions and mixed numbers (e.g., by using visual fraction models or equations to represent the problem). (5.NF.6)
6	**Understand** the concept of a ratio and use ratio language to describe a ratio relationship between two quantities. (6.RP.1)	Fluently divide multi-digit numbers. (6.NS.2) Fluently add, subtract, multiply, and divide multi-digit decimals. (6.NS.3)	Use ratio and rate reasoning to solve real-world and mathematical problems (e.g., by reasoning about tables of equivalent ratios, tape diagrams, double number line diagrams, or equations). (6.RP.3)

Chapter 4

Grade	Conceptual Knowledge Target Standard	Procedural Fluency Target Standard	Application Target Standard
7	**Understand** multiplication is extended from fractions to rational numbers by requiring that operations continue to satisfy the properties of operations, particularly the distributive property, leading to products such as $(-1)(-1) = 1$ and the rules for multiplying signed numbers. (7.NS.2a partial standard)	Fluently add, subtract, multiply, and divide rational numbers. (7.NS.1, 2) Solve multi-step real-life and mathematical problems posed with positive and negative rational numbers in any form. (7.EE.3) Fluently solve one-variable equations of the form $px + q = r$ and $p(x + q) = r$. (7.EE.4)	Understand subtraction of rational numbers as adding the additive inverse, $p - q = p + (-q)$. Show that the distance between two rational numbers on the number line is the absolute value of their difference, and apply this principle in real-world contexts. (7.NS.1c)
8	**Understand** that a function is a rule that assigns to each input exactly one output. The graph of a function is the set of ordered pairs consisting of an input and the corresponding output. (8.F.1)	Solve linear equations with one variable, including examples with infinitely many solutions or no solutions. (8.EE.7) Solve real-world and mathematical problems involving volume of cylinders, cones, and spheres and know the formulas. (8.G.9)	Construct a function to model a linear relationship between two quantities. Determine the rate of change and initial value of the function from a description of a relationship or from two (x, y) values, including reading these from a table or from a graph. Interpret the rate of change and initial value of a linear function in terms of the situation it models, and in terms of its graph or a table of values. (8.F.4)

(California Department of Education, 2015; Adapted from TN Core 2015)

Grade-Level Problem Types for Rigor

We study mathematics in order to make sense of the world in which we live. The demonstration of both procedural and conceptual knowledge is done through real-world applications. This measures a student's ability to interpret the world mathematically, which is the ultimate objective.

Research indicates that problems that target conceptual knowledge highlight a connected web of knowledge, a network in which the discrete information is as prominent as the linking relationships (Hiebert 1986). Problems of this type assess multiple skills. These include the ability to explain judgment (Canobi 2004), explain why procedures work (Berthold and Renkl 2009), and generate or select definitions and concepts (Star and Rittle-Johnson 2009).

Problems that highlight procedural knowledge incorporate a series of steps or actions done to accomplish a goal (Canobi 2009). Procedural problems generally highlight solving problems in a prescriptive manner, and the outcome is measured by accuracy of the answer or steps taken.

All three components of rigor—conceptual understanding, procedural skill, and application—are necessary factors in providing a deep exploration of the content. It is important to note as we begin to explore problem types, a single problem may not target the development of a particular knowledge domain exclusively. A range of problems should be selected that help students develop a comprehensive understanding of a mathematical topic inclusive of all three components of rigor.

This section provides examples of different problems at various grade levels that target the development of each dimension of rigor. The characteristics of these problems are highlighted to help you "calibrate" your lens of rigor. This involves understanding the features of problem types that target a particular domain. You will then be able to not only design specific problem types, but also understand what dimension of rigor is being addressed. It should be noted that these sample problems primarily target a particular domain, but secondary and tertiary emphasis may exist as well.

Characteristics of Conceptual Problems

Conceptual knowledge helps students understand the underlying meaning of topics and how they are connected to overarching big ideas in mathematics. It is often referred to as the "why" of mathematics. The qualities of conceptual problems include:

- Explicit or implicit understanding of the principles that govern the problem
- Understanding of relatedness between ideas in the problem
- Explanation of purpose and utility
- Understanding of the "why" in mathematics
- Connecting seemingly discrete topics to an overarching concept
- Multiple representations of a derived answer

Figure 4.3 shows examples of problems where the primary objective is building conceptual understanding.

Figure 4.3 Examples of Conceptual Knowledge Problems

<table>
<tr><th>Grade</th><th>Conceptual Knowledge</th></tr>
<tr><td>K</td><td>Decompose the number 9 in more than one way. Compare your answer with a partner.
Compare two objects with a measurable attribute in common (e.g. shorter/taller, more/less, longer/shorter).
Identify a shape as two-dimensional or three-dimensional.</td></tr>
<tr><td>1</td><td>If 3 + 9 = 12, then 9 + 3 = 12. Explain why.
How many groups of ten are there in 56? What does the remainder represent?</td></tr>
<tr><td>2</td><td>Compare 312 and 502 using < , > , or =. Write two comparison statements.
Write the number 785 in expanded form.</td></tr>
<tr><td>3</td><td>What does the fraction $\frac{3}{4}$ represent?
The fraction $\frac{5}{2}$ is located between what two numbers on a number line?
Explain why $\frac{1}{2}$ is equivalent to $\frac{3}{6}$.</td></tr>
<tr><td>4</td><td>When 250 is divided by 33, the quotient is 7 with 19 left over. Express this as an equation involving all four numbers.
Compare 0.083 with 0.83.
What is a perpendicular line?</td></tr>
<tr><td>5</td><td>Write 256.39 in expanded form.
Use a model to show 2.3 × 1.6.
What does the point (3, 5) represent in the coordinate plane?</td></tr>
<tr><td>6</td><td>What is a ratio?
Describe the additive and multiplicative structure in the table below.
<table><tr><td>3</td><td>5</td></tr><tr><td>6</td><td>10</td></tr><tr><td>9</td><td>15</td></tr><tr><td>12</td><td>20</td></tr></table></td></tr>
<tr><td>7</td><td>What is the relationship between a proportion and a ratio?
A rectangle is twice as long as it is wide. Find as many different ways as you can to write an expression for the perimeter of such a rectangle.</td></tr>
<tr><td>8</td><td>Explain the Pythagorean Theorem, including its use.
Approximate $\sqrt{2}$ and provide a rationale.</td></tr>
</table>

Characteristics of Procedural Problems

Procedural knowledge helps students use operational skills. It is often referred to as the "how" of mathematics. The qualities of procedural problems include:

- Use of algorithms to calculate answers
- Following prescriptive steps
- Rote memorization without guarantee of true understanding of underlying concepts
- Emphasizing skills, methods, and facts
- Often decontextualized

Figure 4.4 shows examples of problems where the primary objective is building procedural fluency.

Figure 4.4 Examples of Procedural Problems

Grade	Procedural Fluency			
K	$1 + 5 = ?$	$4 + 3 = ?$	Count to 20 by ones.	$4 - 3 = ?$
1	$14 - ? = 8$	45 $+ 18$	$32 - 11 = ?$	
2	$9 + 6 + 8 = ?$	$400 + 60 + 5 = ?$	425 $- 327$	
3	$3 \times 5 = ?$	$4 \times ? = 12$	$24 \div 8 = ?$	
4	$7 \times \frac{2}{5} = ?$	$5\frac{1}{3} + 2\frac{2}{3} = ?$		
5	$(8 + 4) \times 3 = ?$ $4{,}567 \div 34 = ?$	$8 - (\frac{1}{2} + \frac{1}{3}) = ?$ $\frac{7}{8} - \frac{3}{4} = ?$	$34 - 4 \times 5 = ?$	
6	Find the area using the formula $A = \frac{1}{2}bh$, where $b = 4$ in. and $h = 7$ in. Simplify: $3(x + 4) + 2x$ Calculate: $\frac{2}{3} \div \frac{3}{4} = ?$			
7	Solve: $3x + 4 = 12$ Solve the proportion: $\frac{x}{6} = \frac{10}{30}$ Compute: $-20 - (-3) = ?$	Find the quotient: $-345 \div 9 = ?$		
8	Convert $\frac{1}{6}$ to a decimal. Write a numerical expression for $3^2 \times 3^{-3}$.			

Characteristics of Application Problems

Application problems use conceptual knowledge and procedural knowledge to solve problems situated in a real-world context. This is often referred to as the "when" and "where" of mathematics. The qualities of application problems include:

- A real-world context
- Making meaning of a scenario
- Modeling problems that require mathematizing situations
- Demonstrating use of skills and knowledge in an actual setting
- Interdisciplinary connections

Figure 4.5 shows examples of problems where the primary objective is building application skills.

Figure 4.5 Examples of Application Problems

Grade	Application
K	Six apples are on the table. Two are red and the rest are green. How many apples are green? Two kittens sat on the grass. Three more kittens came and sat there. How many kittens are on the grass now?
1	Mike took 5 stickers he no longer wanted and gave them to Susan. Now Mike has 9 stickers remaining. How many stickers did Mike have in the beginning? There are 7 children and some chairs. A child sits in each chair. Two children don't have a chair. How many chairs are there?
2	John had 26 baseball cards. His mom gave him some more cards. Now he has 78 baseball cards. How many baseball cards did his mom give him? Ben had 23 marbles while Julie had 15. They decided to put all of their marbles in a box. How many marbles do they have all together?
3	There are 12 apples on the counter. If you are sharing the apples equally among 3 bags, how many apples will go in each bag? There are five nests in the chicken coop with 3 eggs in each nest. If the farmer wants 25 eggs, how many more eggs does she need?
4	A used bicycle costs $75 while a brand new one costs $300. How many times more expensive is the new bike compared to the old bike? Your class is collecting bottled water for a service project. The goal is to collect 300 bottles of water. On the first day, Max brings in 3 packs with 6 bottles in each pack. Sarah wheels in 6 packs with 6 bottles in each container. About how many bottles of water still need to be collected?
5	Mary is making two types of cookies. She needs $\frac{2}{3}$ of a cup of a sugar for one cookie recipe and $\frac{1}{4}$ of a cup of a sugar for the second cookie recipe. How much sugar does she need all together? Brad had $\frac{3}{4}$ of a pizza left. He gave $\frac{1}{3}$ of the leftover pizza to his sister. How much of the whole pizza did his sister get?

Grade	Application
6	There are 3 brownies to share equally among 5 students. What is the amount of brownie that each student receives? What is the unit rate? A juice recipe calls for 5 cups of orange juice for every 2 cups of cranberry juice. How many cups of orange juice are needed for a batch that uses 8 cups of cranberry juice?
7	Jacob sees a jacket that costs \$32 before a sale. During the sale, prices on all items are reduced by 25%. What is the sale price of the jacket? A youth group is going on a trip to the state fair. The trip costs \$52.75 per person. Included in that price is \$10.25 for a concert ticket and the cost of 3 passes, two for rides and one for the game booths. Each of the passes costs the same price. Write an equation representing the cost of the trip and determine the price of one pass.
8	Three gift cards and 4 movie tickets cost \$188 altogether, while 2 gift cards and 3 movie tickets cost \$116 altogether. Explain how to use this information to find the cost of one gift card and one movie ticket. A car rental company charges \$45 per day for a car as well as a flat fee of \$25 for a car's GPS navigation system. Write an equation for the cost in dollars as a function of the number of days the car is rented. What is the initial value for this function?

Rigor Calibration Tool

Now that you have examined the characteristics and examples of the dimensions of rigor—conceptual understanding, procedural fluency, and application—it is your turn to analyze some examples to identify the dimension(s) of rigor and provide a rationale for your selection. The Rigor Calibration Tool can be used to help you evaluate problems at your own grade level when deciphering the primary and secondary (if applicable) dimension(s) of rigor.

Figure 4.6 Rigor Calibration Tool

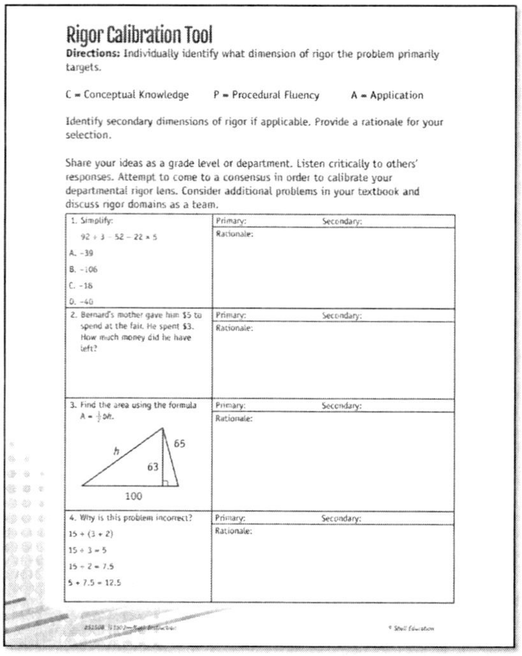

Rigor Calibration Tool

Directions: Individually identify what dimension of rigor the problem primarily targets.

C = Conceptual Knowledge P = Procedural Fluency A = Application

Identify secondary dimensions of rigor if applicable. Provide a rationale for your selection.

Share your ideas as a grade level or department. Listen critically to others' responses. Attempt to come to a consensus in order to calibrate your departmental rigor lens. Consider additional problems in your textbook and discuss rigor domains as a team.

1. Simplify: $92 \div 3 - 52 - 22 \times 5$ A. −39 B. −106 C. −18 D. −40	Primary: Secondary: Rationale:
2. Bernard's mother gave him \$5 to spend at the fair. He spent \$3. How much money did he have left?	Primary: Secondary: Rationale:
3. Find the area using the formula $A = \frac{1}{2}bh$. h, 65, 63, 100	Primary: Secondary: Rationale:
4. Why is this problem incorrect? $15 \div (3 + 2)$ $15 \div 3 = 5$ $15 \div 2 = 7.5$ $5 + 7.5 = 12.5$	Primary: Secondary: Rationale:

You Try It!

1. Download the Rigor Calibration Tool from the Digital Resource CD (filename: calibration.pdf).

2. Review the characteristics and examples of conceptual, procedural, and application problems (page 122–126).

3. Complete the chart by identifying what dimension of rigor the problem primarily targets (conceptual knowledge, procedural fluency, or application), identify a secondary dimension of rigor if applicable, and provide a rationale for your selection.

4. Share your ideas as a grade level or department. Listen critically to others' responses and attempt to come to a consensus. Consider additional problems in your textbook and discuss rigor domains as a team.

Rigor Analysis Tool

Now that you have an understanding of how to identify the dimension(s) of rigor within a problem, the next step is to use that knowledge to make informed decisions about lesson and unit planning. The Rigor Analysis Tool can be used to ensure that there is a range of problem types in a particular unit that develop the desired knowledge domain (conceptual, procedural, application) noted in the unit standards.

The Rigor Analysis Tool can be completed independently or in grade-level teams to determine where the standards explicitly call for conceptual understanding, procedural fluency, and application. By completing the analysis, you will have the opportunity to increase your content knowledge by examining each standard to decipher the primary dimension of rigor being targeted.

Figure 4.7 Rigor Analysis Tool

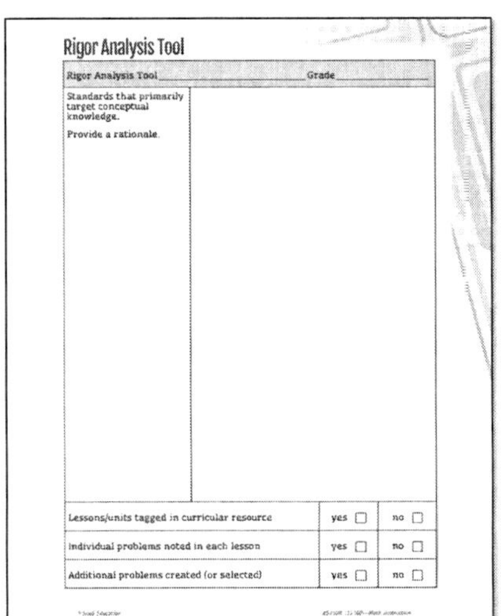

Rigor Analysis Tool

Rigor Analysis Tool______	Grade______
Standards that primarily target conceptual knowledge. Provide a rationale.	

Lessons/units tagged in curricular resource	yes ☐	no ☐
Individual problems noted in each lesson	yes ☐	no ☐
Additional problems created (or selected)	yes ☐	no ☐

You Try It!

1. Download the Rigor Analysis Tool from the Digital Resource CD (filename: rigoranalysis.pdf).
2. Complete the chart by identifying the standards that primarily target conceptual knowledge.
3. Write a brief rationale for each standard.
4. Identify which units/lessons target the identified standards in your curriculum.
5. Tag these units/lessons (e.g., use sticky notes labeled with Concept, Procedure, Application).
6. Strategically select problems in the unit that target the selected dimension of rigor (e.g., conceptual understanding).
7. Create problems yourself, with colleagues, or pull from supplemental resources, as needed.
8. Repeat these steps for standards that primarily target procedural fluency.
9. Repeat these steps for standards that primarily target application.

Rigor Monitoring Tool

The final step in analyzing your curriculum and unit planning for the necessary level of rigor is to use the The Rigor Monitoring Tool as a "rigor barometer" to confirm there is equal balance between conceptual knowledge, procedural fluency, and application. Emphasis is placed on the major clusters, and thus, they are exclusively included in the tool. However, rigor should be applied to supporting clusters (which are aligned to major clusters) as well as additional clusters as needed.

Analyzing your existing curriculum to determine the alignment of rigor to the major clusters produces a roadmap to determine what additional resources must be gathered and/or created to create a rigorous balance.

Use the results from the Rigor Analysis Tool to determine what dimension(s) of rigor a lesson targets, and record the lesson number in the matrix. This can be done to audit an existing curriculum in order to make adjustments or during curriculum development. It is important to remember that over the course of an instructional year, the lessons targeting each dimension of rigor should be balanced for each major cluster. As important as inputting lessons into the chart is the analysis of the holes. This is the final step of the rigor monitoring process. The sample in Figure 4.8 can help you get started.

Figure 4.8 Sample Rigor Monitoring Tool—2nd Grade

Cluster	Conceptual Knowledge	Procedural Skill and Fluency	Application
Represent and solve problems involving addition and subtraction.	2.1–2.3 Problems target conceptual knowledge (2.OA.1) 9.6 Relationship between operations highlighted in only a few problems (2.OA.1)	1.1–1.7 Almost all procedure-based problems (2.OA.1) 2.2–2.7 Primarily procedure-based (2.OA.1) 4.1–4.6 Almost all procedural (2.OA.1) 6.5 & 6.7 Exclusively procedural (2.OA.1)	1.8 Half of the lesson provides application-based problems (2.OA.1) 2.8 Lesson targets a majority of application-based problems (2.OA.1) 7.7 Most of the problems are application-based (2.OA.1)
Add and subtract within 20.		2.2–2.7 Almost all problems (2.OA.2)	2.8 One problem (2.OA.2)

Cluster	Conceptual Knowledge	Procedural Skill and Fluency	Application
Understand place value.	10.1–10.2 Majority of problems discuss why and relevance (2.NBT.1–4)	5.1–5.5 Majority are procedural (2.NBT.1–4)	6.1 & 6.2 Most problems are application-based (2.NBT.2)
Use place value understanding and properties of operations to add and subtract.	3.3 & 3.4 Most problems in these lessons target conceptual understanding (2.NBT.5, 2.NBT.9) 14.5 A few problems are included that discuss the relatedness between operations (2.NBT.5, 9)	2.5–2.7 Majority are procedural problems (2.NBT.5, 2.NBT.9) 6.5–6.7 Most problems are procedural (2.NBT.8–9) 7.1–7.4 Most procedural (2.NBT.5, 2.NBT.8) 9.1–9.4 Most procedural (2.NBT.5, 6, 9) 11.1–11.6 Almost entire chapter procedural, excluding a few application problems (2.NBT.7–9)	5.2–5.3 Most problems are application-based (2.NBT.5) 6.1–6.4 Almost all are application problems (2.NBT.5, 2.NBT.8, 2.NBT.9) 10.2 Real-world problems (2.NBT.8)
Measure and estimate lengths in standard units.		15.1–15.5 Almost all are procedural (2.MD.1–4)	
Relate addition and subtraction to length.	8.3 Relationship problems involving length (2.MD.6)	9.2–9.6 Primarily procedural (2.MD.6) 15.6–7 Primarily procedural (2.MD.5)	

Analysis:

Represent and solve problems involving addition and subtraction

- *Not enough conceptual problems included in the book*
- *Need to find supportive resources for conceptual knowledge development specifically targeting 2.OA.1*

Add and subtract within 20.

- *Book appears to be void of conceptual problems targeting this cluster*
- *Limited application problems for 2.OA.2*

Understand place value.

- *Additional conceptual and application problems needed for 2.NBT.1–4*

Use place value understanding and properties of operations to add and subtract.

- *Majority of problems target procedural knowledge*
- *Need to supplement with additional conceptual knowledge and application-based problems for 2.NBT.5–9*

Measure and estimate lengths in standard units.

- *Book appears to be void of problems targeting conceptual knowledge and application problems (2.MD.1–4)*
- *Additional resources needed for this domain*

Relate addition and subtraction to length.

- *Application problems missing (2.MD.5, 6)*
- *Additional resources needed for application problems (2.MD.5, 6)*
- *Need more lessons that target conceptual knowledge (2.MD.5, 6)*

You Try It!

1. Download the Rigor Monitoring Tool from the Digital Resource CD.
2. Use the results from the Rigor Analysis Tool (pages 127–128) to help you complete the chart.
3. Analyze and record your findings to determine the alignment of rigor.
4. Collaborate with your grade-level colleagues to share your results and gather and/or create additional resources to create a rigorous balance in curriculum.

Now What?

As a result of completing the Rigor Monitoring Tool, if you determine that your curricular resources do not provide a balanced rigorous exploration of the major concepts in your grade level, you may want to consider alternate resources to supplement your curriculum. Additionally, you might want to consider modifying the existing problems in your curricular resource to create the desired balance. Figure 4.9 provides examples of problems across the three domains of rigor. It is important that you understand the relationship between problem types in each domain so that adjustments can be made when planning your lesson. The problems in Figure 4.9 also serves as a model for how to go from one problem type to the next. You may need to create or find new problems depending on identified student deficits relating to a particular domain of rigor.

Also, notice in Figure 4.9 that the standard remains constant throughout the three domains of rigor. No problem is more or less important than the next. All three types help develop a comprehensive and "rigorous" understanding of the standard, which is why you must strive for balance.

Think about the following questions as you analyze the chart:

- What do you notice about the three problems for each standard?
- How are they different? How are they similar?
- How might you modify problems in your curricular resource when needed?

Chapter 4

Figure 4.9 Examples Across the Three Domains of Rigor

Standard	Procedural Problem	Conceptual Problem	Application Problem
K.CC.5	9 + 2 = ____	How does the sum of 9 + 2 compare to each individual number?	Ethan had a bag of lollipops. He gave nine friends one lollipop each. He then gave his two teachers one lollipop each. How many lollipops did Ethan give out? Explain your answer.
1.OA.1.A	$9 + n = 14$	What does the unknown value in $9 + n = 14$ represent? How can you rewrite this math sentence using a different operation?	Nine students were sitting on the bench at the start of recess. Some more students walked over and sat with them during recess. Now, there are 14 students on the bench. How many students sat on the bench during recess? Explain your thinking.
2.NBT.A.1	Consider the number 240. In what place value is the 2?	How many groups of ten are there in the value of 240? How many groups of one hundred?	Sandra needs to pack 240 donuts. Ten donuts can be packed in one box. Ten boxes can be packed into one case. How many boxes are needed? How many cases are needed? Are all of the cases full? Explain your reasoning.
3.OA.3.A	$42 \div 7 = n$	What is the relationship between the dividend of 42, the divisor of 7, and the resulting unknown quotient?	There are 42 desks in Ms. Manzone's classroom. If Ms. Manzone puts 7 desks in each row, how many rows of desks are in the classroom? Justify your answer.
4.NF.4.C	$4 \times \frac{1}{2} = ?$	Use fraction tiles to find the total value of 4 groups of $\frac{1}{2}$. Explain how you know.	Four friends ate lunch together at the park. If each friend ate $\frac{1}{2}$ of a submarine sandwich, how many whole sandwiches did they eat together? Draw a model to explain your thinking.
5.NF.A.1	$\frac{1}{4} + \frac{7}{8} = ?$	How might you determine the sum of $\frac{1}{4}$ and $\frac{7}{8}$ using equivalent fractions? What indicators in the sum point to the value being greater than or equal to one?	Jada is training for her school's summer run competition. She wants to run 1 mile each day to prepare. Jada runs to the mall, which is $\frac{1}{4}$ of a mile from her house. She then runs to the school, which is $\frac{7}{8}$ of mile from the mall. If she stops at the school, will she have trained enough for the day? Explain your reasoning.

Standard	Procedural Problem	Conceptual Problem	Application Problem
6.RP.A.3c	Convert $\frac{17}{20}$ to a percent.	How might you determine if 17 out of 20 is greater than 90%? What do the values of 17 and 20 represent in your response?	Wendy took a math quiz with 20 questions and answered 17 questions correctly. She has to answer 90% of the questions correctly in order to receive an A. Did Wendy answer enough questions correctly to receive an A? Present your argument.
7.EE.B.4b	Solve the inequality. $50 + 10h \geq 120$	Express the mathematical sentence in your own words. $50 + 10h \geq 120$ What does the statement mean? What might be another way to write this sentence and maintain the same meaning?	Corinne is saving money to buy a new phone. The phone costs $120. She received $50 for her birthday. Corinne plans to babysit at a rate of $10 per hour. Write an inequality that represents the least number of hours Corinne needs to babysit in order to earn the minimum amount needed to purchase the phone. Explain how the inequality was derived? Interpret your results.
8.G.B.7	Let a = 90ft., b = 90ft. Find c. $a^2 + b^2 = c^2$	Find the value of the hypotenuse in the isosceles right triangle. 90 ft. 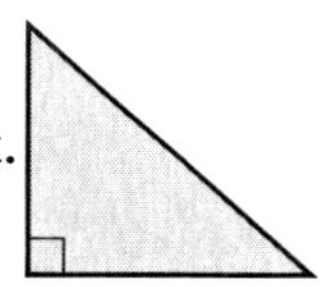 What is the relationship between the legs and the hypotenuse?	In the World Series, the catcher threw out the base runner attempting to steal second base. The distance between home base and first is 90 ft. The distance between each successive base is the same. How far did the catcher have to throw to get the runner out at second base? Justify your position.

Stop and Think

Question 1

How might your understanding of rigor influence the selection of curricular resources?

Question 2

How does the definition of rigor in mathematics compare to commonly held beliefs regarding classroom rigor?

Question 3

What ongoing measures might you take to create a balance among the three domains of rigor in your curriculum?

Chapter 5

Lesson Design

The medium that exists between the teachers and the students is the lesson. Lesson design allows the teacher to manipulate the classroom environment in order to facilitate learning. Ralph W. Tyler (1949), seminal curriculum theorist, asserts that learning experiences must be woven together in an organized and thoughtful way to form a coherent program. This chapter highlights how to incorporate the instructional shifts and knowledge of the standards (both content and practice) into meaningful units of instruction. The instructional shifts are operationalized in each lesson, targeting student mastery of the standards. In addition, the major criteria required for the creation of effective Common Core experiences is addressed.

- Focus on high-leverage content
- Coherent connections among concepts
- Rigorous exploration of topics
- Mathematical practices intertwined with content
- Range of multiple models of teaching

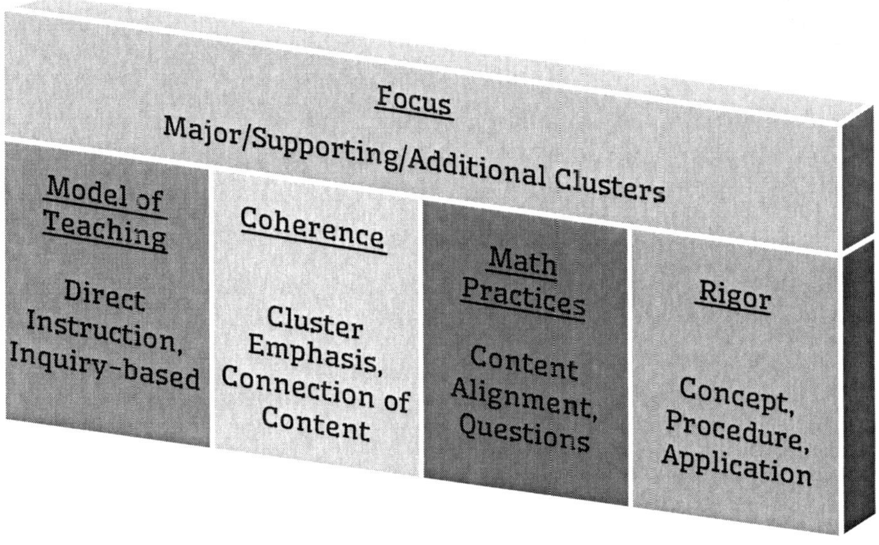

The instructional shifts and the mathematical practices have been introduced in previous chapters of this book. The primary point of emphasis in this section will be models of teaching in a unit of study.

The following resources are provided:

- **Focus-Coherence-Rigor (FCR) Unit Frame**—This is a comprehensive framework that connects the instructional shifts in an overarching unit guide. Connections are made between clusters, standards, and models of teaching. Individual lessons are situated in this larger unit frame.
- **Inquiry-Based Lessons**—Investigative inquiry-based models primarily target the development of conceptual knowledge. Resources are provided to help you construct your own lessons. Exemplars are included.
- **Direct Instruction Lessons**—Direct instruction models primarily target the development of procedural knowledge. With the advent of the Common Core, an emphasis on inquiry does not eliminate the need for direct instruction. Resources are provided to help you construct your own lessons. Exemplars are included.
- **Unit Checklist**—How closely aligned are your existing curricular resources with the Common Core shifts and expectations? A unit checklist has been provided to determine the readiness of your curriculum in order to make necessary adjustments.
- **Instructional Strategies**—Instructional strategies are embedded to increase a student's comprehension of the math topic under investigation. Examples of instructional strategies and models are provided for each grade level. A template is included to help teachers apply the strategies presented at their appropriate grade level.

Focus-Coherence-Rigor (FCR) Unit Frame

When designing lessons, they must be situated in an overarching unit of study. Once the overall unit concepts are established, the individual lessons within the unit should be developed. The FCR Unit Frame is used to explicitly show the connection between the clusters, standards, instructional shifts, and selected models of teaching. It is a road map that can be used to highlight relatedness between the lessons and the overarching cluster concepts as well as the relatedness between lessons themselves.

FCR Unit Frame Protocol

Step 1: Identify the desired cluster for the unit. The unit of instruction centers around the identified cluster(s). Once the cluster(s) is identified, determine if the cluster is a major, supporting, or additional cluster. This knowledge will be used to determine time allocation. If the cluster is a supporting cluster, indicate the major cluster that is being supported (see Chapter 2).

Step 2: Identify the targeted standards within the cluster(s) that will be addressed throughout the unit. Examine the standard and select the appropriate mathematical practices that are most closely aligned (see Chapter 1).

Step 3: Individual lessons can now be outlined. Establish the learning objectives (LO): students will be able to... (SWBAT).

Step 4: Incorporate elements into each lesson that show the relatedness between the lesson and the cluster. Use the cluster connect box to record the script or strategy that will be used (see Chapter 3).

Step 5: Incorporate elements into each lesson that show the relatedness between lessons. Use the lesson connect box to incorporate these instructional strategies (see Chapter 3).

Step 6: Based on your established learning objective, choose a corresponding model of teaching (direct instruction, inquiry-based investigation, etc.). More detail is provided on models of teaching (MOT) following the FCR Unit Frame. Each selected model of teaching will serve as the individual frame to construct the daily lesson.

Step 7: Record the math practice(s) that will be targeted during the lesson.

Step 8: Determine which component(s) of rigor will be highlighted and the associated problems in the note section (see Chapter 4).

A sample is provided to help you get started.

Figure 5.1 Sample FCR Unit Frame

Cluster(s): *Extend understanding of fraction equivalence and ordering*

Emphasis: (M) S A

Build fractions from unit fractions

Emphasis: (M) S A

Standard:

4.NF.1 Explain why $\frac{a}{b}$ is equivalent to a fraction $\frac{(n \times a)}{(n \times b)}$ by using visual fraction models

Math Practices	
1	2
3	4
5	6
7	8

Standard:

4.NF.2 Compare two fractions with different numerators and different denominators.

Math Practices	
1	2
3	4
5	6
7	8

Standard:

4.NF.3 Understand a fraction $\frac{a}{b}$ with a > 1 as a sum of fractions $\frac{1}{b}$

Math Practices	
1	2
3	4
5	6
7	8

Cluster Connect
Discuss Ordering

Cluster Connect
How have you extended your understanding of fraction equivalence?

LO: *SWBAT explain equivalent fractions.*
MOT: *Inquiry-based Model*
MPs: *7*
Rigor Emphasis: *Application*
Note (pg. # created): *Performance Task*

Lesson Connect
How might you use a model to see the number and size of fractional parts?

LO: *SWBAT use fraction models to explain equivalence*
MOT: *Direct instruction Model*
MPs: *7, 8*
Rigor Emphasis: *Procedure, Concept*
Note (pg. # created): *#2-12 pg. 169*

Lesson Connect
How might you use this knowledge to solve real-world problems?

Cluster Connect
How does ordering help in the real world?

LO: *SWBAT explain equivalent fractions using fraction models.*
MOT: *Direct Instruction Model*
MPs: *2*
Rigor Emphasis: *Application*
Note (pg. # created): *#30, 36, 42, 49 pg. 171*

Lesson Connect
Not all fractions have the same denominator. How do you compare fractions with different denominators?

Cluster Connect
Fractions with different denominators can be ordered.

LO: *SWBAT compare fractions with different denominators.*
MOT: *Inquiry-Based Model*
MPs: *2, 7*
Rigor Emphasis: *Concept, Application*
Note (pg. # created): *#46, pg. 174*

Lesson Connect
How might you compare fractions to common benchmark fractions?

Cluster Connect
What patterns do you notice when fractions are ordered.

LO: *SWBAT compare fractions to benchmark fractions.*
MOT: *Direct Instruction Model*
MPs: *3, 7*
Rigor Emphasis: *Concept*
Note (pg. # created): *Choose key benchmark $\frac{1}{2}$, $\frac{1}{4}$*

Lesson Connect
Not all fractions have the same denominator. How do you compare fractions with different denominators?

Cluster Connect
How are fractions with numerators > 1 composed of unit fractions?

LO: *SWBAT understand a fraction as a sum of unit fractions.*
MOT: *Direct Instruction Model*
MPs: *3, 7*
Rigor Emphasis: *Prodcedure, Concept*
Note (pg. # created): *insert SBAC/PARCC problems.*

Lesson Connect
How might you decompose a fraction in different ways?

1. Download the FCR Unit Frame Protocol Template from the Digital Resource CD (filename: fcr.pdf).
2. Review the FCR Unit Frame Protocol sample (pages 138–140).
3. Follow the protocol to complete the chart.
4. Use the FCR Unit Frame to help you create individual lesson plans and a summative assessment for the unit of study.

You have now created your unit road map that will ultimately lead to a summative unit assessment (see Chapter 6). Individual lessons must now be constructed. Models of teaching may be used to design these lessons.

Models of Teaching

Models of teaching describe prescriptive steps, grounded in learning theory that can be used to design lesson experiences for students. Historically, the seemingly default model in mathematics was that of direct instruction. An emphasis has now been placed on incorporating more investigative inquiry-based models into mathematics instruction. Figure 5.2 presents multiple models of instruction. This does not represent an exhaustive list of teaching models. Educators are encouraged to learn more about the models listed here as well as other models supported by theoretical research. This resource focuses on two models of teaching:

- direct instruction
- inquiry-based group investigation

Direct instruction is a model of teaching that consists of the teacher explaining the new skill or concept to all students and then allowing time for students to practice in a structured manner. The inquiry-based model of teaching consists of students organized into problem-solving groups and investigating new concepts, in which the teacher acts as a consulting guide through the exploration. These two represent the current dominant models in mathematics.

Figure 5.2 Models of Teaching

Rigor and Models of Teaching			
Conceptual Knowledge	**Information Processing Learning Theory**	**Concept Attainment Model**	• Examples and nonexamples of a concept are presented—two pre-determined classifications • Students determine categorization and defining attributes • Rules and concepts are defined
		Concept Formation Model	• Many objects and items are provided and students must classify • Students explain grouping and rationale
		Advanced Organizer Model	• Organizer is presented that clarifies the aims of the lessons • Connects student's existing knowledge to organizer using known context • New material presented connecting to organizer
	Socio/Constructivist Learning Theory	**Inquiry-Based Model**	• New scenario (puzzlement) presented • Students problem solve and test developed conjectures in organized groups • Analyze results and recycle for continued exploration
		Role-Playing Model	• Used for counseling to deal with human relations problems • Systematic program of social education • Exploration of beliefs and values
Procedural Knowledge	**Behavioral Learning Theory**	**Direct Instruction Model**	• Explicit demonstration of a skill or concept by the teacher • Structured, guided, and independent practice for the student
		Simulation Model	• Broad topic incorporated into real-world simulation activity • Roles and responsibilities assigned to students • Analyze results and relate to the course of study

(Adapted from Joyce, Weil, and Calhoun 2004)

As seen in Figure 5.2, certain models of teaching are advantageous to use when targeting a particular knowledge domain (conceptual knowledge development vs. procedural knowledge development).

The question is often asked, "When do we use a particular model?" While there is no hard and steady rule, the use of an inquiry-based model to start a unit (and to start the major sub-topics within a unit) is recommended due to the opportunity provided to explore the unknown elements of a new topic. The direct instruction model can be used thereafter once a topic is introduced to teach a procedural skill to apply the new concept. Using one model all of the time does not provide an opportunity for meaningful learning across all knowledge domains (factual, conceptual, procedural, and metacognitve). When planning a lesson, it is important to consider the learning objective emphasis when determining which model of teaching may be more appropriate. Learning objectives include a targeted thinking skill and the concept being studied.

Think of a lever or seesaw with the skill being on one side (e.g., cognitive level, Bloom's, critical-thinking skill) and the mathematics topic on the other side (e.g., polygons or ratios). If the weight of the lesson is on developing the student's ability to use the thinking skill (e.g., to calculate), a direct instruction model may be most useful due to its emphasis on a learned behavior. If the emphasis of the lesson is on understanding the mathematical concept (e.g., concept of place value or equations), then it may be best to use an inquiry-based model that promotes enduring learning in a social context. Regardless of whether the lesson is intended for first-time, primary, whole-group instruction or tier 2 or tier 3 intervention, it is important to determine the point of emphasis in the lesson based on the needs of the students as well as determine an appropriate model of teaching. However, it is important to be mindful and understand that there is no hard and steady rule that dictates a model of teaching selection.

Inquiry-Based Model

When students have an opportunity to investigate a concept through inquiry and creative exploration, the likelihood that they retain understanding of that topic increases. To meet the expectations of today's standards, units of study often begin with a concept-building lesson. Procedures are then taught in context. This does not suggest that this sequence must always be used. Providing students an opportunity to negotiate a mathematics scenario in the absence of direct instruction of new skills may create anxiety on the part of the student. It is important to discuss the purpose of investigative inquiry-based lessons with students to reduce anxiety and to encourage perseverance through the task. Inquiry-based lessons serve as diagnostic tools and not summative assessments.

Multiple inquiry-based lessons should be used throughout a unit of study. Your current curriculum resource can be used to create investigative inquiry-based lessons.

The following is an example of an inquiry-based lesson template:

Figure 5.3 Inquiry-Based Math Lesson Template

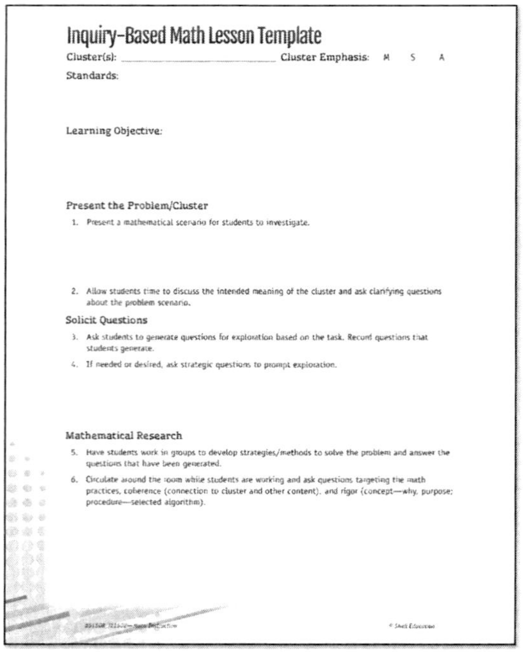

Inquiry-Based Math Lesson Template

Cluster(s): ______________________ Cluster Emphasis: M S A

Standards:

Learning Objective:

Present the Problem/Cluster

1. Present a mathematical scenario for students to investigate.
2. Allow students time to discuss the intended meaning of the cluster and ask clarifying questions about the problem scenario.

Solicit Questions

3. Ask students to generate questions for exploration based on the task. Record questions that students generate.
4. If needed or desired, ask strategic questions to prompt exploration.

Mathematical Research

5. Have students work in groups to develop strategies/methods to solve the problem and answer the questions that have been generated.
6. Circulate around the room while students are working and ask questions targeting the math practices, coherence (connection to cluster and other content), and rigor (concept—why, purpose; procedure—selected algorithm).

Meaningful Mathematics (M²)

What is central to an inquiry-based lesson is the selection of a problem that is deemed worthy of investigation. These are problems that give students an opportunity to learn mathematics in a meaningful way. The M² criteria chart may be used to vet a selected math problem.

The Meaningful Mathematics (M²) criteria chart describes characteristics of math problems that ensure that students are engaged in high levels of cognitive demand. M² problems provide varied avenues for creative thinking to occur. It is not realistic to think that all M² problems will have all of these characteristics. However, selected problems should have an overwhelming majority of these characteristics. Grade-level teams can discuss modifications to existing problems to further align problems with these criteria. Creation of problems should also be encouraged.

Figure 5.4 Meaningful Mathematics Chart

Meaningful Mathematics (M²) Criteria Chart

- The solution path is nonstandard and not predictable
- Requires an analysis of constraints that may lead to qualifying statements
- Creates a productive struggle due to the non-routine structure of the problem
- Targets multiple dimensions of rigor
- Encourages student to determine relatedness between task and cluster
- Requires synthesis of existing knowledge in order to create knowledge needed for task
- Provides real-world context

A full-sized version of the Meaningful Mathematics Chart can be located on the Digital Resource CD as a resource to be printed and displayed in your classroom.

Inquiry-Based Lesson

To begin, the teacher presents a mathematical scenario for investigation. For example, "A welder has an $8\frac{1}{4}$ in. strip of metal that she is cutting into $1\frac{3}{8}$ in. pieces. How many pieces will she make?" Then, the teacher solicits questions from students regarding the task. One way to do this is to provide time for students to individually record one or two questions and then have students share their questions in groups. Each group is then responsible for selecting a representative question that will then be recorded on the board. Possible student questions might include:

- Will there be any material leftover?
- What strategy will I use to determine how fourths are related to eighths?

If needed, the teacher can also incorporate targeted questions for exploration such as the following:

- How would you describe the problem in your own words?
- What do you know from the previous unit that may apply to this problem?
- What is known and unknown?

The teacher circulates around the room and asks these targeted questions based on students' responses.

It is important that teachers do not provide answers or over-scaffold their assistance during an inquiry-based task. Resources should be available for use, but teachers should not promote the use of any particular materials. As previously shared, the lesson should be developed around a problem that has been vetted and found worthy of investigation.

While walking through the classroom, teachers should be mindful of certain pathways that they will highlight to unveil salient mathematical concepts. These are often concepts that have been noted in advance when the teacher solved the problem multiple ways prior to the lesson. However, it is not limited to these predetermined pathways. Through exploration, novel ideas and solution pathways are often created and should be shared. It is also important to note that the teacher should vary the order that pathways are presented (e.g., concrete to abstract vs. abstract to concrete). The teacher should avoid always selecting a concrete example such as an illustration used to solve the problem as the first pathway being shared versus the selection of an abstract algebraic approach that solely uses decontextualized symbols. There is no hard and set rule other than varying the pathway presentation order so that students are not conditioned to think that one way is preferred over the other way. For example, in this particular problem one student may show a drawing that has the $8\frac{1}{4}$ in. strip measured and partitioned into 6 smaller sections that are $1\frac{3}{8}$ in. each (concrete). Another student may show the quotient of $8\frac{1}{4}$ and $1\frac{3}{8}$ using division procedures.

As the solution pathways are presented, students act as critics. Careful preparation should be made for students who argue and critique. Students should always be aware that they are playing one of these two roles. Questions they ask themselves and others are presented in Figure 5.5. An enlarged version of the Perpetual Inquiry Chart can be found on the Digital Resource CD as a resource to be printed and displayed in your classroom.

Figure 5.5 Perpetual Inquiry Chart

Perpetual Inquiry

Arguer

Did I state my claim?

What type of supportive evidence is provided?

How strongly does the evidence align to my claim?

Am I incorporating the language of the discipline?

Do the statements build upon each other in a logical and coherent manner?

Critic

Do I note any ambiguities?

Are there any assumptions?

Am I judging against established criteria?

Was relevance discussed?

Does the argument consider all cases?

Is there an exception to the argument presented?

Teachers may ask the following questions while students are sharing and summarizing to help guide and focus the discussion.

- What do the solution pathways have in common? Any differences?
- Do you agree or disagree? Why?
- What is the relationship between this concept and the previous concept, multiplication of fractions?

The teacher may act as the moderator of the exchange of conversation between the student presenting his/her argument and the remaining students responding critically. Even though a group investigation lesson covering new content is not to be used as a summative assessment, the teacher should correct any glaring misconceptions and preview the skills and knowledge that will be learned in the unit (or lessons) that follows.

The lesson can end by visiting the overarching cluster and asking students reflective questions.

- What meaning has been made as a result of this lesson?
- What are some of the major topics that we will explore in this upcoming unit?
- How does this lesson align with the overarching idea of the cluster?

As students are encouraged to use creative-thinking skills and innovate, so should educators as you build your own inquiry-based lessons.

1. Download the Inquiry-Based Lesson template from the Digital Resource CD (filename: inquirybase.pdf).
2. Review the inquiry-based model and sample lessons (pages 145–155).
3. Use the Inquiry-Based Lesson Template to create your own lesson.

The following are inquiry-based lesson examples.

Lesson Sample: *Sticky Situation*

Grades K–2

Cluster(s):

Understand place value.

Emphasis: (Major) [circled]
Supporting
Additional

Standards:

- Compare two three-digit numbers based on meanings of the hundreds, tens, and ones digits, using >, =, and < symbols to record the results of comparisons. (2.NBT.4)
- Attend to precision. (MP6)
- Look for and make use of structure. (MP7)

Learning Objective:

Students will be able to compare numbers using place value meaning.

Present the Problem/Cluster

1. Present the following mathematical scenario and cluster for students to investigate. Say, "Today you will use what you know about place value to solve problems."
   John and Maria each have stickers that come in different packs. Pack A stickers contain 100 stickers in each pack. Pack B stickers contain 10 stickers in each pack. Pack C stickers contain one sticker per pack. John has 3 pack A stickers, 5 pack B stickers, and 17 pack C stickers. Maria has 2 packs of A stickers, 15 packs of B stickers, and 18 packs of C stickers. Who has the most stickers? Explain your reasoning by drawing pictures, writing number sentences using >, =, and < symbols."
2. Allow students time to discuss the intended meaning of the cluster and ask clarifying questions about the problem scenario. A question students may ask would be: How will place value help me solve this problem?

Solicit Questions

3. Ask students to generate questions for exploration based on the task. Record questions that students generate. Some questions may be:
   - How do the tens compare to the hundreds in our total number of stickers?
   - Did I answer all parts of the problem?
   - Are there other ways to solve this problem?
   - Is it possible to open the pack of stickers and group them differently?
   - Why does Pack C only have one sticker?
4. If needed or desired, ask the following strategic questions to prompt exploration:
   - What mathematical language can you use to explain ______? (MP6)
   - Have you considered all parts of the problem in your argument? (MP6)
   - What patterns do you notice? (MP7)

Lesson Sample: *Sticky Situation* (cont.)

Grades K–2

Mathematical Research

5. Have students work independently, then in groups to develop strategies/methods to solve the problem and answer the questions that have been generated. A second option involves students immediately working in groups. Examples of strategies students may use include:
    - Drawing individual packs and then regrouping the packs of ten and the packs of one
    - Using repeated addition
6. Circulate around the room while students are working and ask questions targeting the math practices, coherence, and rigor. Note selected solution pathways that will be shared. Suggested questions:
    - What is the relationship between the number of stickers in each of the three packs? (MP6)
    - How does place value impact your solution pathway? (Coherence)
    - What mathematical operations are you using to make sense of the problem? (Rigor)
    - How many mathematical sentences can you construct using the <, > , = symbols? (MP7)

Share and Summarize

7. Allow students to present their findings from research and their argument. Consider selecting students who illustrate specific solution pathways and solutions that are connected to the solicited questions from earlier in the lesson. For example:
    - Maria has more stickers. Even though she has a smaller number of Pack A, she has a larger combined number of Packs B and C. Therefore, she has more stickers.
8. Allow students to act as respectful critics of each other's work by asking the following questions at various points throughout the presentations.
    - What evidence do you have that shows that is true?
    - I disagree because _______.
    - Did you consider the sizes of each pack?
9. Select solution pathways and connect to solicited questions in multiple ways, including:
    - How are the solution pathways similar/different?
    - How does this solution emphasize the use of place value?
    - What questions from the board have we answered?

Lesson Sample: *Sticky Situation* (cont.)

Grades K–2

Recycle and Forecast

10. Highlight the studied cluster from the lesson and allow students to discuss the meaning that was created. Students can generate additional questions as a result of the research. Examples include:
    - What other ways can I regroup numbers to find a total sum?
    - How might multiplication be used here?
11. Ask students new questions as a result of the exploration and provide questions that will be explored in future lessons. Some examples include:
    - What place value strategies can you use to compare numbers?
    - How do the ones, tens, and hundreds compare to each other?
12. Close the lesson with a journal prompt. Say, "How does this lesson increase your understanding of place value? Record your response in your journal."

NOTES

Lesson Sample: *Full of Pie*

Grades 3–5

Cluster(s):

Develop understanding of fractions as numbers.

Emphasis: (Major) Supporting Additional

Standards:

- Explain equivalence of fractions in special cases, and compare fractions by reasoning about their size. (3.NF.3)
- Make sense of problems and persevere in solving them. (MP1)
- Construct viable arguments and critique the reasoning of others. (MP3)
- Look for and make use of structure. (MP7)

Learning Objective:

Students will be able to compare fractions to determine equivalence.

Present the Problem/Cluster

1. Present the following mathematical scenario and cluster for students to investigate. Say, "We will investigate a mathematical scenario to further shape our understanding of fractions as numbers. I want us to think about the following question: When are fractions equivalent?"
2. Allow students time to think independently, then say, "Today, we will explore the concept of equivalent fractions. We are also going to continue to develop our skills as mathematicians—to notice the structure of fractions and to construct viable arguments to defend our solutions. Group A has three pies that will be shared evenly among four students. Group B has two pies that will be shared evenly among three different students. If you want to have the most pie, which group would you join? Present your argument and explain your thinking."
3. Students will then ask clarifying questions. Some examples may be:
   - How does the structure of a fraction help me understand how to approach the problem?
   - How are fractions equivalent if they do not look alike?

Solicit Questions

4. Ask students to generate questions for exploration based on the task. Record questions that students generate. Some questions may be:
   - Are the pies the same size?
   - How can I compare three pieces to four pieces mathematically? I can see that the slice of pie in Group B is larger.
   - Does the pie have to be cut into only thirds or fourths?
5. If needed or desired, ask the following strategic questions to prompt exploration:
   - How did you select a strategy to try? (MP3)
   - What evidence is used to support your position? (MP3)
   - What patterns do you notice? (MP7)

Lesson Sample: *Full of Pie* (cont.)

Grades 3–5

Mathematical Research

6. Have students work independently, then in groups to develop strategies/methods to solve the problem and answer the questions that have been generated. A second option involves students immediately working in groups. Examples of strategies students may use include:
    - Students partition pies in multiple ways that may allow them to compare both pies.
    - Students partition pies into thirds and fourths and compare sizes.
7. Circulate around the room while students are working and ask questions targeting the math practices, coherence (connection to cluster and other content), and rigor (concept—why, purpose; procedure—selected algorithm). Note selected solution pathways that will be shared. Suggested questions:
    - What is the relationship between the number of pies and the fraction produced? (Analytical Questions)
    - How does your understanding of fractions help find a solution? (Coherence)
    - What mathematical operations are you using to make sense of the problem? (Rigor)
    - What parts of the problem are you using to compare one group with the next? (MP7)

Share and Summarize

8. Allow students to present their findings from research and their argument. Consider selecting students who illustrate specific solution pathways and solutions that are connected to the solicited questions from earlier in the lesson. For example:
    - I would join Group B because it appears that my slices of pie are larger than those from Group A. Also there are fewer people sharing the pies in Group B, so I will have more of the pie. They are not cut into the same number of pieces so I have to estimate.
9. Allow students to act as respectful critics of each other's work by asking the following questions at various points throughout the presentations.
    - You are assuming that you have more based on the size of the pieces. How do you prove this with evidence?
    - I believe that each person in Group A has more pie because I have sliced each pie using the same number of slices to compare.
10. Select solution pathways and connect to solicited questions in multiple ways, including:
    - How are the solution pathways similar/different?
    - How does this solution emphasize the concept of fraction equivalence?
    - What questions from the board have we answered?

Lesson Sample: *Full of Pie* (cont.)

Grades 3–5

Recycle and Forecast

11. Highlight the studied cluster from the lesson and allow students to discuss the meaning that was created. Students can generate additional questions as a result of the research. Examples include:
    - It is not always easy to cut the pie or any shape into the same number of pieces in order to compare. How might we compare fractions with different denominators using some type of rule?
    - How are common unit fractions helpful in comparing problems?
12. Ask students new questions as a result of the exploration and provide questions that will be explored in future lessons. Some examples include:
    - How might you compare different parts or fractions that refer to the same whole?
13. Close the lesson with a journal prompt. Say, "How does this lesson increase your understanding of equivalent fractions? Record your response in your journal."

NOTES

Lesson Sample: *All Gas-ed Up*

Grades 6–8

Cluster(s):

Use place value understanding and properties of operations to add and subtract.

Emphasis: (Major)
Supporting
Additional

Standards:

- Use ratio and rate reasoning to solve real-world and mathematical problems, e.g., by reasoning about tables of equivalent ratios, tape diagrams, double number line diagrams, or equations (6.RP.3)
- Construct viable arguments and critique the reasoning of others (MP3)
- Look for and make use of structure. (MP7)

Learning Objective:

Students will be able to use ratios to construct an argument.

Present the Problem/Cluster

1. Present the following mathematical scenario and cluster for students to investigate. Say, "We will investigate a mathematical concept to understand how ratios can be used in the real world. I want us to think about the following question: When have you used a ratio to understand a situation at home or with friends?"
2. Allow students time to discuss the intended meaning of the cluster and ask clarifying questions about the problem scenario. Some examples may include:
   - If I am comparing two things, does that mean I have a ratio?
   - Does it matter which item comes first in the ratio?
3. Present a mathematical scenario for investigation. Say to the class, "Today we will explore the concept of ratio reasoning. We are also going to continue to develop our skills as mathematicians—to determine tools needed in a math situation and to construct viable arguments to defend our solutions. A store owner offers 10 gas points for every $50 you spend in the store. For every 10 gas points you earn, you get to take $0.10 off every gallon of gas you buy. Currently, gas is $3.50 a gallon. When Mike fills up his car, he usually puts 12 gallons of gas in it. How much money must Mike spend at the store in order to fill up his car for less than $35.00?"

Lesson Sample: *All Gas-ed Up* (cont.)

Grades 6–8

Solicit Questions

4. Ask students to generate questions for exploration based on the task. Record questions that students generate. Some examples may be:
   - How much does it cost to fill the tank with no discount?
   - What ratios are going to be helpful in solving this problem?
   - How much does the gas have to cost per gallon in order to stay under $35.00?
5. If needed or desired, ask the following strategic questions to prompt exploration:
   - What mathematical tools can be used to help visualize the scenario? (MP3)
   - How did you select a strategy to try? (MP3)
   - What evidence is used to support your position? (MP3)

Mathematical Research

6. Have students work independently, then in groups to develop strategies/methods to solve the problem and answer the questions that have been generated. A second option involves students immediately working in groups.
7. Circulate around the room while students are working and ask questions targeting the math practices, coherence (connection to cluster and other content), and rigor (concept—why, purpose; procedure—selected algorithm). Suggested questions:
   - What is the relationship between the savings needed and the $50? (Analytical Questions)
   - How does your understanding of ratios help find a solution? (Coherence)
   - What mathematical operations and algorithms are you using to make sense of the problem? (MP1)

Share and Summarize

8. Allow students to present their findings from research and their argument. Consider selecting students who illustrate specific solution pathways and solutions that are connected to the solicited questions from earlier in the lesson.
9. Allow students to act as respectful critics of each other's work by asking questions at various point throughout the presentations. Connect solution pathways to the students' solicited questions in multiple ways, including:
    - How are the solution pathways similar/different?
    - How does this solution emphasize the use of ratios to solve problems?
    - What questions from the board have we answered?

Recycle and Forecast

10. Solicit new questions as a result of the exploration and provide questions that will be answered in future lessons. Examples:
    - How are unit rates useful in solving problems of this type?
    - How does the concept of a percent relate to this scenario?
11. Ask students, "How does this lesson increase your understanding of ratios? Record your response in your journal."
12. Record any additional questions that students generate as a result of the research. Example:
    - How might we use ratios and equations to solve problems?

NOTES

Direct Instruction Model

As mentioned in the beginning of the chapter, there are multiple models of teaching. An inquiry-based group investigation lends itself to the development of conceptual knowledge as students explore with minimal scaffolding. The idea that a rigorous, standards-based lesson should not involve direct instruction is a misconception. There are established algorithms and procedures in the discipline of mathematics that can be taught using a direct instruction model of teaching. Students are encouraged to extend ideas taught and to find multiple ways to solve problems. It should be noted that a direct instruction model does not mean that the teacher leads the instruction for the duration of the lesson.

The syntax of a direct instruction model is as follows. The *T* (teacher) and the *S* (student) represent the dominance of the teacher and students during each phase. The teacher transitions from assuming all responsibility for completing the task (T) while the student (s) listens, to the student assuming full responsibility for independently completing the task (S) while the teacher monitors and gives feedback (т). This transition is shown in the model that follows.

Phase I: Orientation/Motivation (Ts): The teacher establishes a framework and introduces the new content.

1. Introduce activities that elicit students' relevant existing knowledge
2. Discuss the objective of the lesson
3. Provide an overview of the lesson

Phase II: Demonstration (T): The teacher demonstrates explicitly while students listen and take notes.

4. Model skill or content
5. Present material in small steps
6. Provide varied examples of the skill or content
7. Re-explain difficult concepts

Phase III: Check for Understanding (TS): The teacher facilitates discussion with students.

8. Ask a range of questions (factual, analytical, evaluative)
9. Provide feedback
10. Correct misconceptions

Phase IV: Structured Practice (TS): The teacher and students work together to solve problems.

11. Provide varied examples and representations of the skill or content (symbolic, pictorial, written)
12. Reinforce accurate responses
13. Highlight connections to objective and cluster

Phase V: Guided Practice (тS): The students work in pairs/groups.

14. Students work through multiple problem types collaboratively
15. Teacher monitors student work and provides corrective feedback

Phase VI: Independent Practice (S): The students work independently.

16. Purpose is to reinforce learning to increase retention and fluency
17. Homework may serve as independent practice
18. Students practice without assistance—delayed feedback

A Direct Instruction Model Template can be found on the Digital Resource CD. The following are sample direct instruction lessons.

Lesson Sample: *Sums of Fruit*

Grades K–2

Cluster(s):

Understand and apply properties of operations and the relationship between addition and subtraction.

Emphasis: (Major)
Supporting
Additional

Standards:

- Apply properties of operations as strategies to add and subtract. Examples: If 8 + 3 = 11 is known, then 3 + 8 = 11 is also known. (Commutative property of addition.) To add 2 + 6 + 4, the second two numbers can be added to make a ten, so 2 + 6 + 4 = 2 + 10 = 12. (Associative property of addition.) (1.OA.3)
- Make sense of problems and persevere in solving them. (MP1)
- Look for and make use of structure. (MP7)

Learning Objective:

Students will be able to apply operation strategies to add numbers.

Build Motivation

1. Post on the board the following: Math Cluster, Learning Objective, MP Prompts. Discuss the concept of "order" when adding, and ask, "Does the order matter when adding numbers?" Then, review the learning objective and the cluster.
2. Let students ask clarifying questions. Examples include:
   - How do we know which number to add first?
   - Does the order matter when we add?
   - What does it mean to persevere?
3. Allow students time to discuss the intended meaning of the cluster as it relates to previous lessons and the upcoming lesson. Examples include:
   - The cluster means that there are rules that we use when we add and subtract numbers.

Demonstrate

3. Present the following problem, saying, "Mary has four apples, eight oranges, and six bananas. How many pieces of fruit does Mary have? Show how you can figure out the sum using at least two different ways." Options would include:
   - 4 + 8 + 6 = ? (Add them in the order presented)
   - 4 + 6 + 8 = ? (First add four and six, then add eight)
4. Highlight the prompt and answer questions, such as:
   - What do I notice about the sum when I add numbers in any order? What does this mean about addition?
   - I notice that the sum is the same no matter what order the numbers are added. I also notice that I can add different parts of a sentence to determine the value of the whole sentence. Finding a sum of ten in the sentence is an operation strategy.
5. Students should listen actively and generate questions and comments. Examples include:
   - Why did you only switch the 6 and the 8?
   - Are we able to switch any of the numbers when we are adding?

Lesson Sample: *Sums of Fruit* (cont.)

Grades K–2

Check Understanding

6. Ask students questions such as:
   - What dimension of rigor does this problem target?
   - What meaning is made from this problem regarding addition?
   - What is the difference between adding 2 + 5 versus 5 + 2?
   - What are some operation strategies that can be used when adding numbers?
7. Field questions recorded by students during the demo phase, and clarify misconceptions, such as:
   - Students may attempt to apply this thinking to subtraction.
   - Students may confuse the two terms commutative and associative.
   - Students may think that grouping of terms being added only works in limited cases involving two terms or three as the examples shown.
8. Students ask clarifying questions. Examples would include:
   - What is rigor?
   - When we add, the order does not matter. Is this always true?
9. Have students work with a partner to discuss the concepts and generate additional ideas, which could include:
   - I can add numbers together in any order.
   - I may look for numbers that add up to values like five and ten to help me add.

Structured Practice

10. Collaborate with students to explore additional problems, while reinforcing accurate response and highlighting the cluster. Help students solve the following problems:
    - 2 + 7 + 8 = 10 + _____
    - 9 + 2 = _____ + 9
    - 2 + 3 + 4 = _____ + 7
    - 10 + 6 = _____ + 4 + 6
    - _____ + 7 = 7 + 3
    - 8 + 5 + _____ = 10 + 8
11. Students collaborate to solve the problems, and record any questions they may have.

Lesson Sample: *Sums of Fruit* (cont.)

Grades K–2

Guided Practice

12. Monitor students' work and provide corrective feedback. Consider rigor domains when assigning the following problems:
    - Find the sum of the digits using as many ways as possible. Provide an argument. (3, 6, 7)
    - Record how many buttons, zippers, and shoelaces the students in your group are wearing. Use these values to create two different number sentences showing the total.
13. Assign the "Coherence Ladder" activity (see Chapter 3)
14. Have students work in groups of two or three to solve problems, and explain the connection between the lesson and the overarching cluster. Examples include:
    - There are rules that allow us to add numbers in any order.
    - Do these rules work for subtraction, too?

Independent Practice

15. Provide a task to reinforce learning and increase retention. Examples include:
    - Write the missing number that would make each equation true. Provide an argument.

      4 + 7 + 3 = 10 + ______

      2 + 1 + 6 = ______ + 6

      ______ 8 = 8 + 3

      5 + 2 = ______ + 5

      10 + 9 = ______ 3 + 7

      3 + 5 + ______ = 10 + 5
    - Determine whether or not this operation strategy works for subtraction. Be prepared to share your argument.
16. Students practice on their own without assistance.

NOTES

Lesson Sample: *Cookie Bake Sale*

Grades 3–5

Cluster(s):

Gain familiarity with factors and multiples.

Emphasis: (Major) — circled

Supporting

Additional

Standards:

- Find all factor pairs for a whole number in the range 1–100. Recognize that a whole number is a multiple of each of its factors. Determine whether a given whole number in the range 1–100 is a multiple of a given one-digit number. Determine whether a given whole number in the range 1–100 is prime or composite. (4.OA.4)
- Reason abstractly and quantitatively. (MP2)
- Look for and make use of structure. (MP7)

Learning Objective:

Students will be able to determine the relatedness between whole numbers and their corresponding factors.

Build Motivation

1. Post on the board the following: Math Cluster, Learning Objective, MP Prompts. Discuss the concept of multi-digit multiplication and the relationship with place value. (Example: $6 \times 7 = 42$ versus $6 \times 70 = 420$ versus $6 \times 700 = 4{,}200$.) Introduce the concepts of factors and multiples using the multi-digit example above. Say, "Today we will examine the relationship between factors and multiples and classifying numbers as prime or composite."
2. Let students ask clarifying questions. Examples include:
   - Is this similar to writing a number in an expanded form?
   - Patterns of ten can be useful in this example.
3. Allow students time to discuss the intended meaning of the cluster as it relates to previous lessons and the upcoming lesson. Examples include:
   - What is the connection between a factor and a multiple?
   - Why is it important to understand how to find multiples?
   - What is a composite number?

Lesson Sample: *Cookie Bake Sale* (cont.)

Grades 3–5

Demonstrate

3. Present the following problem on multiples of 3:
   - 3 = 3 × 1   12 = 3 × 4
   - 6 = 3 × 2   15 = 3 × 5
   - 9 = 3 × 3   18 = 3 × 6
   - The products 3, 6, 9, 12, 15, 18 are multiples of 3. The product results from multiplying factors. The factors are the numbers being multiplied.
     - Example: Three and one are the factors of 3. Three and five are factors of 15.
   - If a number has only two unique factors, 1 and itself, then it is prime. If the number has more than two factors, then it is composite.
     - Example: Six is a composite number because 6 = 3 × 2 and 6 = 6 × 1. Six has more than two unique factors.
     - Example: Three is a prime number because 3 = 3 × 1. Three has only two unique factors.
4. Highlight the prompt and answer questions and comments, such as:
   - What do I notice about the characteristics of the numbers (whole numbers, positive numbers)?
   - I notice that the factors are multiplied to form a product. The product is a multiple of each factor. Each product is either prime or composite based on the parts. The structure of the problem dictates the category.
5. Students should listen actively and generate questions and comments. Examples include:
   - Are factors used to make multiples?
   - Do they act as opposite functions?
   - How many multiples does a number have?

Check Understanding

6. Ask students questions such as:
   - What dimension of rigor does this problem target?
   - What is the relationship between the factors of a number and its classification of prime or composite?
7. Field questions recorded by students during the demo phase, and clarify misconceptions, such as:
   - Students may confuse factors of a number with multiples of a number.
   - Students may not consider all of the factors of a number (e.g., the number and one).
8. Students ask clarifying questions. Examples would include:
   - What are the parts of rigor?
   - Is it possible for a number to be both prime and composite?
9. Have students work with a partner to discuss the concepts and generate additional questions.

Lesson Sample: *Cookie Bake Sale* (cont.)

Grades 3–5

Structured Practice

10. Collaborate with students to explore additional problems, while reinforcing accurate responses and highlighting the cluster. Help students solve the following problem:
    - Lauryn and Corinne made 72 cookies for a bake sale. They will put an equal number of cookies into bags. Lauryn and Corinne want to put more than 2 cookies but fewer than 10 cookies into each bag.
    - Lauryn thinks there are more ways to put an equal number of cookies into bags. Who is correct? Provide an argument.
11. Students collaborate to solve the problems, and record any questions they may have. Examples include:
    - Do we only have a certain number of bags available?
    - Are the cookies the same size?

Guided Practice

12. Monitor students' work and provide corrective feedback. Consider rigor domains when assigning the following problem:
    - Look at the highway signs below. Determine whether each highway sign is a prime or composite number. Provide an argument that includes all possible factors.

13. Answer questions students constructed during the structured practice phase, and provide differentiated support. Examples include:
    - English Language Learners: Teacher provides sentence frames for students to participate in the discussion. (e.g., This number is _____ because there are more than two _____ .
    - Advanced Learners: Depth and complexity—Generate a rule that applies to all prime number factorizations and composite number factorizations.
14. Have students work in groups of two or three to solve problems and explain the connection between the lesson and the overarching cluster. Examples include:
    - The cluster says for us to gain familiarity with factors and multiples.
    - This lesson talks about numbers.

Lesson Sample: *Scooter Adventure*

Grades 6–8

Cluster(s):

Solve real-life and mathematical problems using numerical and algebraic expressions and equations.

Emphasis: Major (circled)
Supporting
Additional

Standards:

- Use variables to represent quantities in a real-world or mathematical problem, and construct simple equations and inequalities to solve problems by reasoning about the quantities. (7.EE.4)
- Reason abstractly and quantitatively. (MP2)
- Look for and make use of structure. (MP7)

Learning Objective:

Students will be able to construct an equation that can be used to solve a real-life math problem.

Build Motivation

1. Post on the board the following: Math Cluster, Learning Objective, MP Prompts. Discuss the concept of numerical expressions and the idea of "solving" a problem. Ask, "How can an expression be used to 'solve' a problem? How does this relate to our cluster? Today we will examine how an expression leads to an equation that can be used to solve a real-world problem."
2. Let students ask clarifying questions. Examples include:
   - How are expressions different from equations?
   - Does a math symbol have to be included in an expression?
3. Allow students time to discuss the intended meaning of the cluster as it relates to previous lessons and the upcoming lesson. Examples include:
   - We are going to be solving problems in the real world by writing expressions.

Demonstrate

4. Present the following problem and demonstrate the application of constructing an expression and equation and using them to solve the problem. Say, "Midtown Bikes rents motorized scooters to tourists for the day. Each rental has a flat fee of $50 per day and a fee of 75 cents per mile. Write an equation to show the total fee based on the number of miles traveled in a day. If the total daily rental cost is $140, how many miles were traveled?"
5. Establish expressions and then determine which have the same value and set them equal to each other. Examples include:
   - Expression 1: $0.75x$; x equals total miles traveled.
   - Expression 2: $0.75x + 50$; cost of mileage traveled plus flat fee.
6. Highlight the prompt and answer questions, such as:
   - What is known and unknown? How can I represent the quantities symbolically or algebraically?
   - What do I notice about the quantities?
7. Students should listen actively and generate questions and comments. Examples include:
   - Are you always adding the flat fee? Why is it not being multiplied by x?
   - How did you convert to get the same units?

Check Understanding

8. Ask students questions such as:
    - What dimension of rigor does this problem target?
    - What is the relationship between the expressions and the eventual equation used?
    - What solution pathway was used to solve the problem?
9. Field questions recorded by students during the demo phase, and clarify misconceptions, such as:
    - Students may confuse the coefficient of the variable with the constant.
    - Students may not convert to equivalent units of measurement (cents and dollars).
10. Students ask clarifying questions. Examples would include:
    - Are expressions contained in equations?
11. Have students work with a partner to discuss the concepts and generate additional questions.

Structured Practice

12. Collaborate with students to explore additional problems, while reinforcing accurate responses and highlighting the cluster. Help students solve problems. (Note: Notice the structure of each problem. What properties are used to solve? How does this concept build upon the previous concept of writing expressions?)
    - Determine the value of x.
        a. $45 = \frac{3}{8}x - 57$
        b. $4(x - 5) = 21$
    - What is the relevance of the equal sign in a problem?
13. Students collaborate to solve the problems and record any questions they may have. Examples include:
    - How can we isolate x in each equation?
    - How can the order of operations help me solve this problem?

Lesson Sample: *Scooter Adventure* (cont.)

Grades 6–8

Guided Practice

14. Monitor students' work and provide corrective feedback. Consider rigor domains when assigning the following problems:
    - English Language Learners: Provide language support that includes a graphic organizer to understand similarities and differences between equations and expressions.
    - Advanced Learners: Provide extension questions such as: How might the mileage rental rate change if the flat fee goes up to $100 and the rider rides the same number of miles to produce a total charge of $140?
15. Assign the "Coherence Ladder" activity (see Chapter 3)
16. Have students work in groups of two or three to solve problems and explain the connection between the lesson and the overarching cluster.

Independent Practice

17. Provide a task to reinforce learning and increase retention. Examples include:
    - Provide at least two different ways that you could solve it.
    - How might this problem be converted into an inequality?
18. Students practice on their own without assistance.

NOTES

1. Download the Direct Instruction Lesson template from the Digital Resource CD (filename: directinstruct.pdf).
2. Review the direct instruction model and sample lessons (pages 156–165).
3. Use the Direct Instruction Lesson Template to create your own lesson.
4. After teaching your lesson, reflect on ways you can improve upon planning or teaching the lesson to help students meet the learning objective.

Common Core Instructional Strategies

Within any given unit of study or a specific lesson, instructional strategies are embedded to increase a student's comprehension of the math topic under investigation. Several strategies are referenced in the Common Core State Standards (2010) ranging from the use of concrete models to pictorial drawings to decontextualized abstract representations. The combination of contextualized and decontextualized strategies increases a student's conceptual understanding of mathematics, resulting in the ultimate goal—application of mathematics in the real world.

Standard for Mathematical Practice 2—Reason abstractly and quantitatively—emphasizes the importance of a student's ability to apply abstract mathematical representations and symbols from a concrete situation and the converse, to be able to describe a concrete context given an abstract representation. It is known that students' knowledge is strongest when they can connect real-world situations, manipulatives, pictures, and spoken and written symbols (Lesh 1990).

Meaningful learning occurs when students are able to organize new concepts in their long-term memory by making connections between familiar concepts. Concrete, everyday representations often serve as these familiar concepts and act as anchors for new concepts.

Curricular resources will often suggest prescribed steps when exploring the relationship between the concrete, pictorial representation, and abstract ideas.

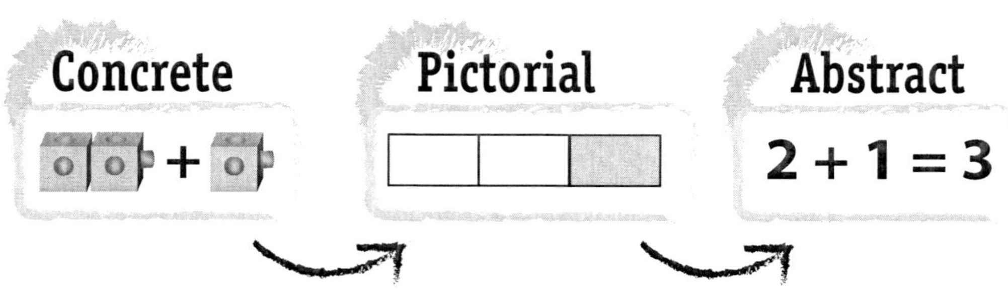

However, it is important to note that students should have the opportunity to not only explore concrete to abstract pathways but also abstract to concrete pathways.

Scientific-based research has not shown what modes of presentations are crucial and what sequence of representations should be used before abstract symbols are introduced (Baroody 1989; Clements 1999). Teachers should be careful about blindly adhering to a concrete-pictorial-abstract sequence, especially when more than one way of thinking about a concrete context exists.

Mathematical Models vs. The Modeling Process

There is a distinctive difference between the use of concrete models to understand mathematical concepts and the process of mathematical modeling. A mathematical model is an isolated representation of a concept. It may be a physical construct or mathematical symbols that represent the concept. A mathematical model is a subset of the process of mathematical modeling.

Mathematical modeling is a cyclical process in which students develop ideas and conjectures about a particular real-world mathematics scenario. They then create a mathematical model that best represents the concept. This model is then tested and refined in an iterative fashion. The results are then explained and justified in the context of the mathematics scenario. The mathematical model is only a component of the entire process of modeling.

Common Core explicitly mentions different instructional strategies, some of which are concrete models that may be applied relating to unique standards. The figures that follow highlight several of these strategies, many of which are instructional models. Though these instructional strategies are listed for a particular grade, they are advantageous to use across grade levels based on students' readiness level and responsiveness to the strategy being implemented.

Figure 5.6 Mathematical Strategies K–8

Kindergarten	
Standard	K.CC.6
Instructional Strategy	Matching and Counting

One-to-one correspondence is used to pair elements from each set. Students will determine which elements are not matched and therefore can identify which set has more or less elements. Students may also count each set and compare the values to determine which set has more or less.

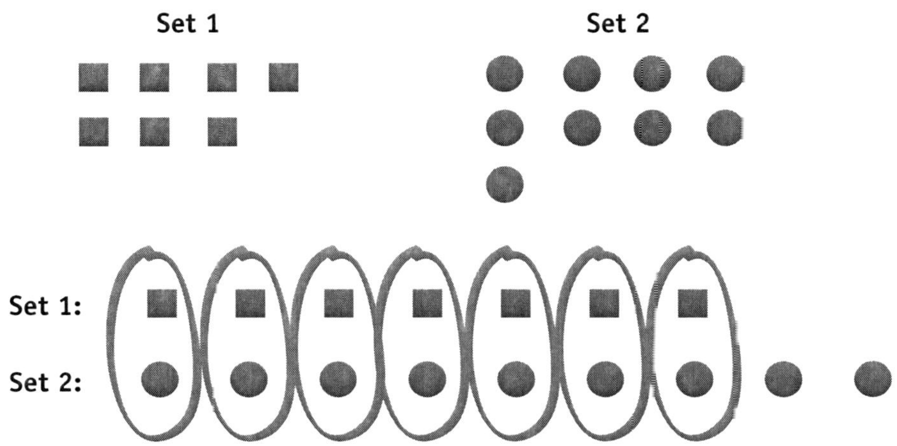

Set 2 has two elements that are not matched. Therefore, set 2 has more elements.

Standard(s)	K.OA.1, K.OA.2, K.OA.3, K.OA.4, K.NBT.1
Instructional Strategy	Use of objects, fingers, mental images, drawings, and sounds (e.g., claps). Use of a ten frame or a number bond.

Students use objects and drawings to determine addition and subtraction problems that involve putting together, adding to, taking apart, and taking from.

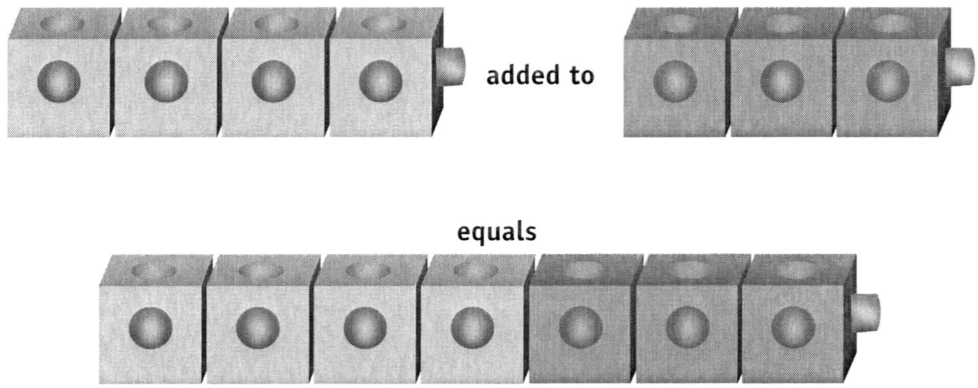

Students place objects on a ten frame (physical one or drawing) and determine how many more are needed to compose ten. Taking away elements from a full ten frame can be used to show how 10 is decomposed into two numbers.

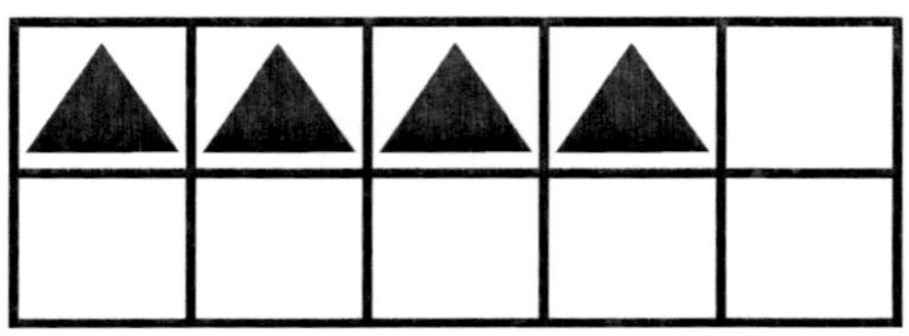

Six additional triangles are needed to make 10.

Number bonds can be used to show the relationship between addition and subtraction and the relatedness between a whole number and its parts.

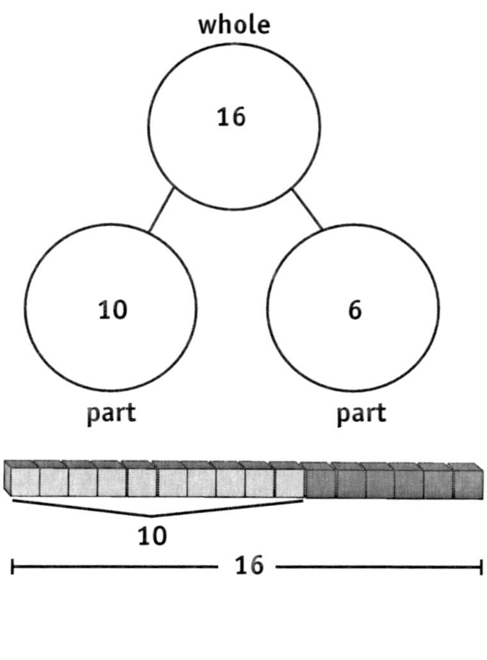

10 + 6 = 16
6 + 10 = 16
16 − 10 = 6
16 − 6 = 10

Standard(s)	K.G.5
Instructional Strategy	Model shapes in the world by building shapes from components (e.g., sticks and clay balls) and drawing shapes.

Students may use clay to build three-dimensional real-world objects. Students can mold clay into a sphere to represent Earth.

Students may use craft sticks to build two-dimensional real-world objects. Students can build triangles, squares, and rectangles. Then, students can count the sides and vertices (corners) of each shape.

First Grade

Standard(s)	1.OA.1, 1.OA.2
Instructional Strategy	Use objects, drawings, and equations with a symbol for the unknown number to represent the problem.

Symbols for the unknown can be used, which later will be represented with variables. Comparison bars can also be drawn.

Elaine has 9 pencils. Jared has 6 pencils. How many more pencils does Elaine have than Jared?

$$6 + \square = 9$$

<table>
<tr><td colspan="2">9</td></tr>
<tr><td>6</td><td>?</td></tr>
</table>

Standard(s)	1.OA.5, 1.OA.6
Instructional Strategy	Count on, make ten, decompose a number leading to a ten, create equivalent but easier or known sums.

Counting on strategies to make 10.

$$7 + 5 = ?$$

$$\underbrace{7 + 3}_{} + 2 = 12$$

$$10 + 2 = 12$$

Decomposing a number leading to 10.

$$16 - 9 = ?$$

$$\underbrace{16 - 6}_{} - 3 = ?$$

$$10 - 3 = 7$$

Standard(s)	1.NBT.4, 1.NBT.6
Instructional Strategy	Use concrete models or drawings based on place value, properties of operations, and/or the relationship between addition and subtraction.

Students may use base-10 blocks or counters.

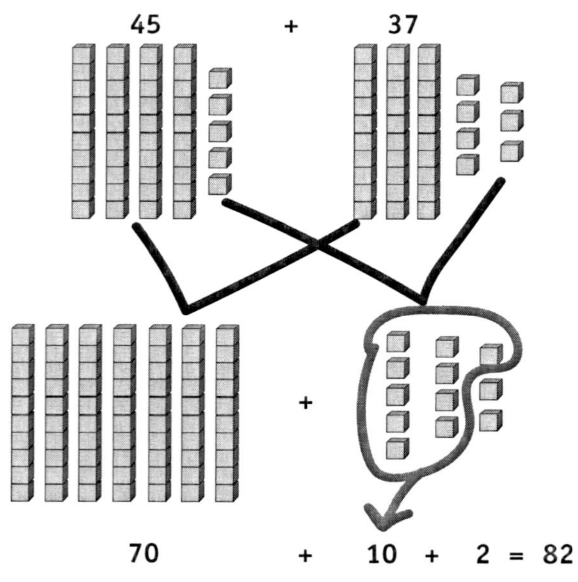

Standard(s)	1.MD.2
Instructional Strategy	Lay multiple copies of a shorter object (the length unit) end to end.

Students use actual unit lengths (centimeters or inches) to determine the length of a larger object. This can be done with unit cubes or unit tape. Objects being measured should be limited to provide no gaps or overlap.

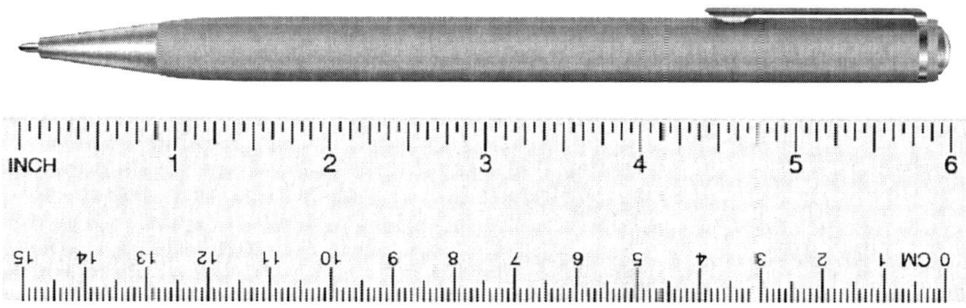

Second Grade

Standard(s)	2.OA.1
Instructional Strategy	Use drawings and equations with a symbol for the unknown number to represent the problem.

Some fans were waiting to get into the basketball game. Six fans decided to leave. There were 25 fans left waiting. How many fans were there before any fans left?

$$\square - 6 = 25$$

The total is written first and then broken apart into the 6 and 25.

Standard(s)	2.OA.3
Instructional Strategy	Pair objects or count them by 2s.

Students pair objects to determine if quantities are even or odd (up to 20).

Determine if 15 is even or odd.

There is one star not paired, therefore 15 is odd.

Standard(s)	2.NBT.5, 2.NBT.7
Instructional Strategy	Use concrete models or drawings based on place value, properties of operations, and/or the relationship between addition and subtraction.

Students use drawings or base-10 blocks.

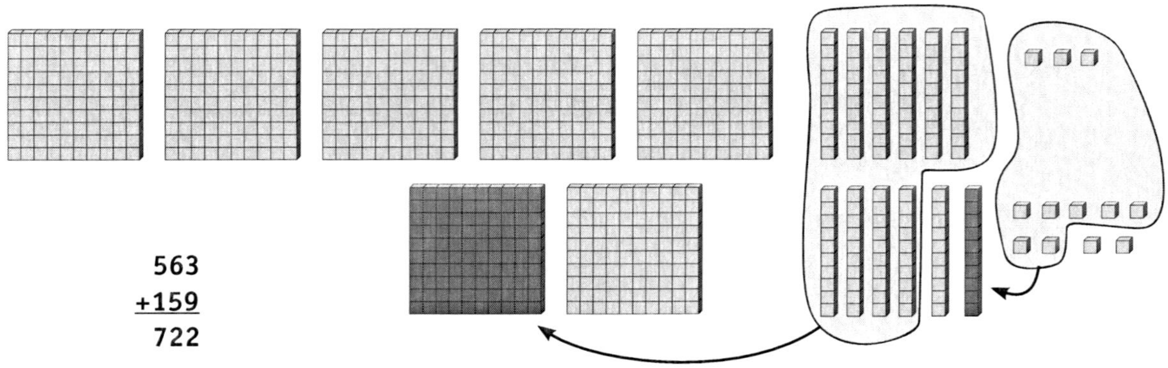

```
  563
 +159
  722
```

Standard(s)	2.MD.5
Instructional Strategy	Use drawings (such as drawings of rulers) and equations with a symbol for the unknown number to represent the problem.

Mark has a kite that is 27 inches tall. Sara has a kite that is 43 inches tall. How many inches taller is Sara's kite?

$$27 + \square = 43$$

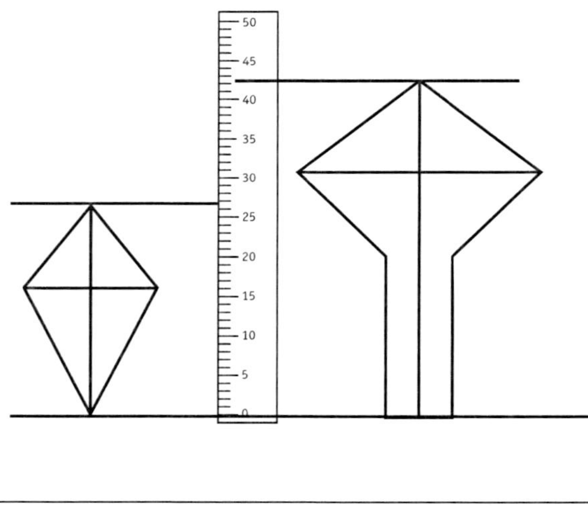

Standard(s)	2.MD.9
Instructional Strategy	Make a line plot, where the horizontal scale is marked off in whole-number units.

Students use rulers to measure and approximate to the nearest whole number. Data is collected on a line plot.

How many inches is your foot? How does the length of your foot compare to your classmates?

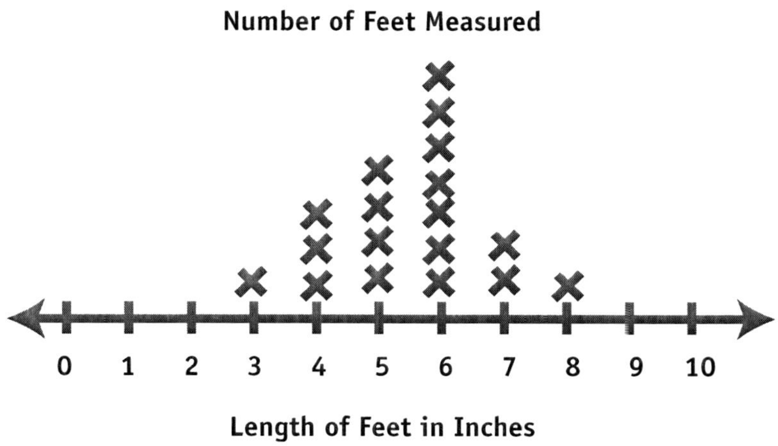

Third Grade

Standard(s)	3.OA.3
Instructional Strategy	Use drawings and equations with a symbol for the unknown number to represent the problem.

Eighteen cupcakes are placed in three boxes with the same amount in each box. How many cupcakes are in each box?

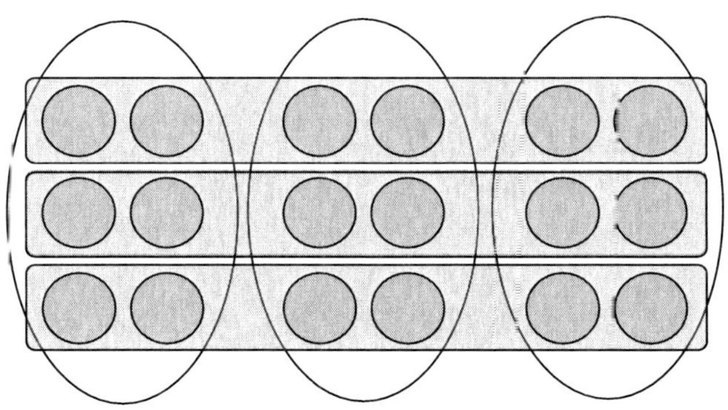

18 ÷ 3 = □

Standard(s)	3.OA.7
Instructional Strategy	Use strategies such as the relationship between multiplication and division (e.g., knowing that 8 × 5 = 40, one knows 40 ÷ 5 = 8) or properties of operations.

Students can turn a division problem into an unknown factor problem. Knowing the related multiplication facts can help a student solve a division problem.

Thirty players were divided into 6 teams.

Think: 6 teams of how many on each team equals 30 players?

$$6 \times \square = 30$$

$$6 \times 5 = 30$$

$$\text{so,}\quad 30 \div 6 = 5$$

Students may also use the Properties of Operations Strategy.

$$7 \times 8 = ?$$

7 × 8 is 7 × 7 (49) plus one more group of 7, which equals 56.

Standard(s)	3.NBT.2, 3.NBT.3
Instructional Strategy	Use strategies and algorithms based on place value, properties of operations, and/or the relationship between addition and subtraction.

Students understand the relationship between 8 × 40 and 8 × 4. If 8 × 4 is 32, then 8 × 40 has to be 10 times as much as 32, since 40 is ten times as much as 4.

Students can apply grouping strategies with place value.

6 x 40

6 groups of 4 tens — or — 24 tens (6 x 4)

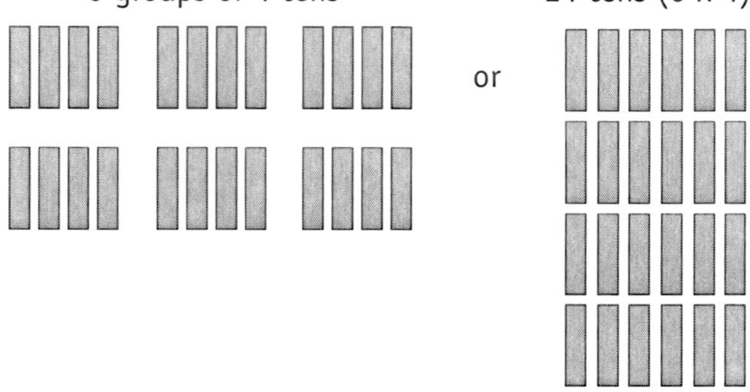

Standard(s)	3.NF.2, 3.NF.3c, 3.MD.1
Instructional Strategy	Use a number line diagram.

$\frac{3}{4}$ on a number line is represented by marking off 3 lengths of $\frac{1}{4}$ from 0

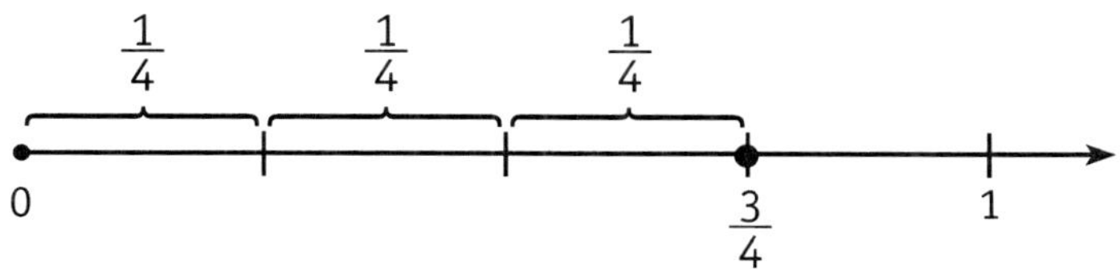

$\frac{5}{3}$ on a number line is represented by marking off lengths of $\frac{1}{3}$ from 0

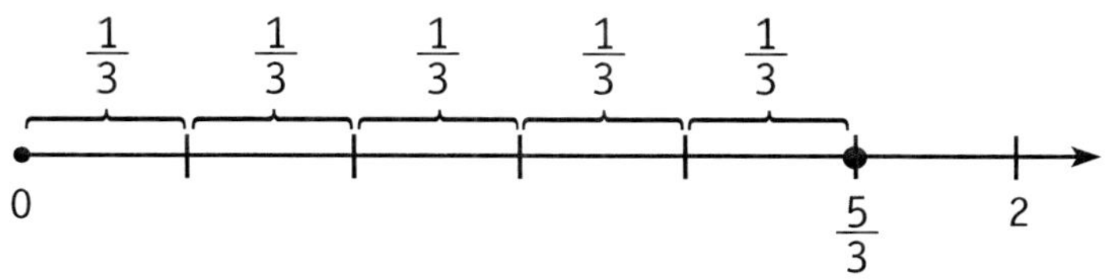

Standard(s)	3.NF.3b, 3.NF.3d
Instructional Strategy	Use a visual fraction model (fraction bar or number line).

$\frac{3}{6}$ is equivalent to $\frac{1}{2}$

Fraction Model

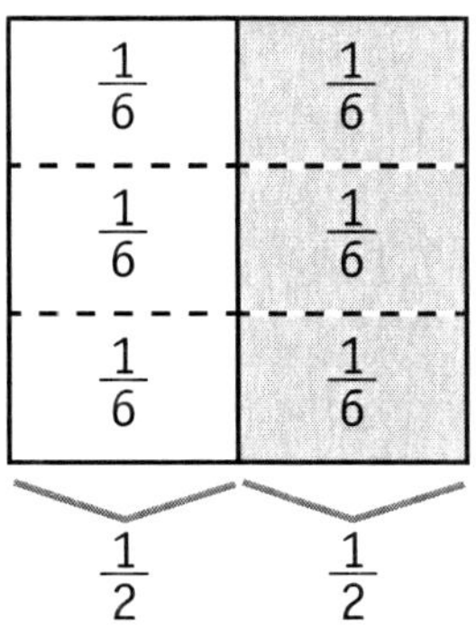

Fraction Bars

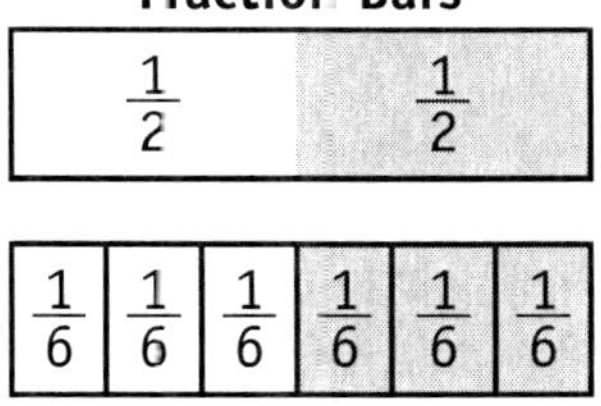

Fraction Number Line

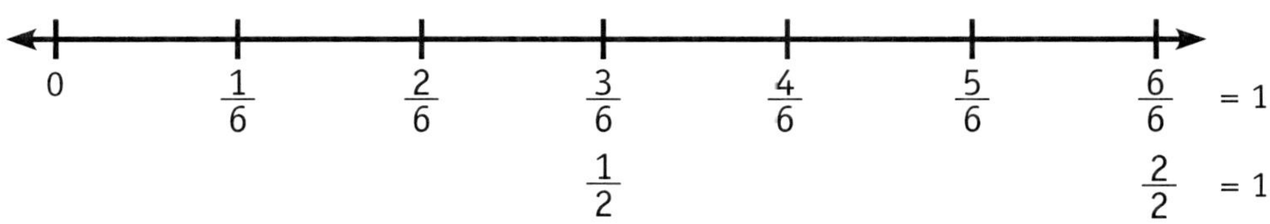

Standard(s)	3.MD.3
Instructional Strategy	Draw a scaled picture graph or a scaled bar graph.

Students might draw a pictograph with symbols that represent multiple units.

Number of Baseball Hits

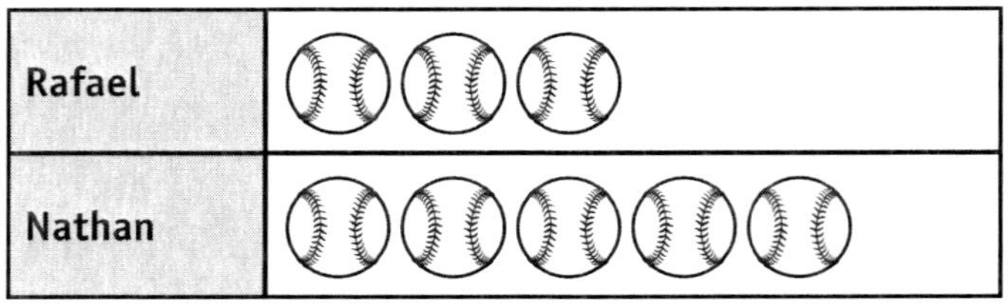

Each (baseball) represents 3 hits.

Standard(s)	3.MD.7a, 3.MD.7c
Instructional Strategy	Find area by tiling; use of area models.

Students may use area models to show the Distributive Property of Multiplication and to find the total area of a figure.

What is the area of a rectangle that is 6 units by 12 units?

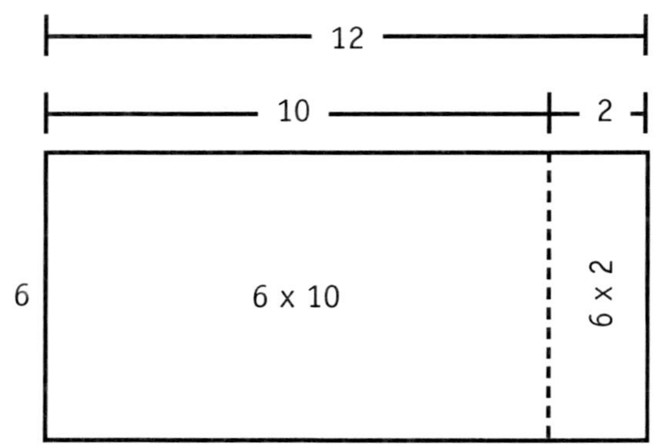

60 + 12 = 72 square units

Fourth Grade	
Standard(s)	4.OA.2
Instructional Strategy	Use drawings and equations with a symbol for the unknown number to represent the problem, distinguishing multiplicative comparison from additive comparison. Use of tape diagrams.

Tape diagrams are rectangular models that compare parts in a problem.

Elizabeth sold 36 tickets to the school music concert. This is 4 times as many as the number of tickets that Sandy sold. How many tickets did Sandy sell?

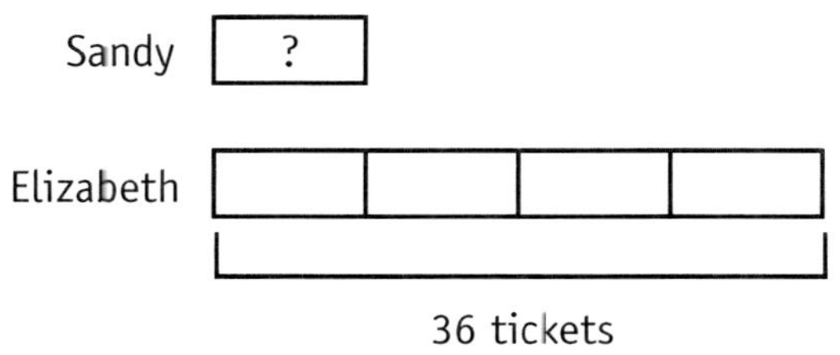

4 x □ = 36

multiplicative comparison

□ + □ + □ + □ = 36

additive comparison

The number of tickets Elizabeth sold is represented by 4 pieces of tape. One piece of tape represents the number of tickets Sandy sold, which is 9.

Standard(s)	4.NBT.5, 4.NBT.6
Instructional Strategy	Illustrate and explain the calculation by using equations, rectangular arrays, and/or area models.

23 × 46 = ?

Students may use a rectangular model. Tens and ones are represented by areas within the model.

```
   46
 x 23
   18 → 3 x 6
  120 → 3 x 4 tens
  120 → 2 tens x 6
+ 800 → 2 tens x 4 tens
1,058
```

46 = 40 + 6; 23 = 20 + 3

	40	6
20	20 x 40 2 tens x 4 tens 8 hundreds = 800	20 x 6 2 tens x 6 12 tens = 120
3	3 x 40 3 x 4 tens 12 tens = 120	3 x 6 = 18

Distributive Property

$$23 \times 46 = (20 + 3) \times 46 = (20 \times 46) + (3 \times 46)$$
$$= 20 \times (40 + 6) + 3 \times (40 + 6)$$
$$= (20 \times 40) + (20 \times 6) + (3 \times 40) + (3 \times 6)$$
$$= 800 + 120 + 120 + 18 = 1{,}058$$

Standard(s)	4.NF.1, 4.NF.2, 4.NF.3b, 4.NF.3d, 4.NF.4a, 4.NF.4b, 4.NF.4c
Instructional Strategy	Use visual fraction models, with attention to how the number and size of the parts differ even though the two fractions themselves are the same size.

Students may use fraction models to represent equivalent fractions with different part sizes.

$$\frac{3}{4} = \frac{9}{12}$$

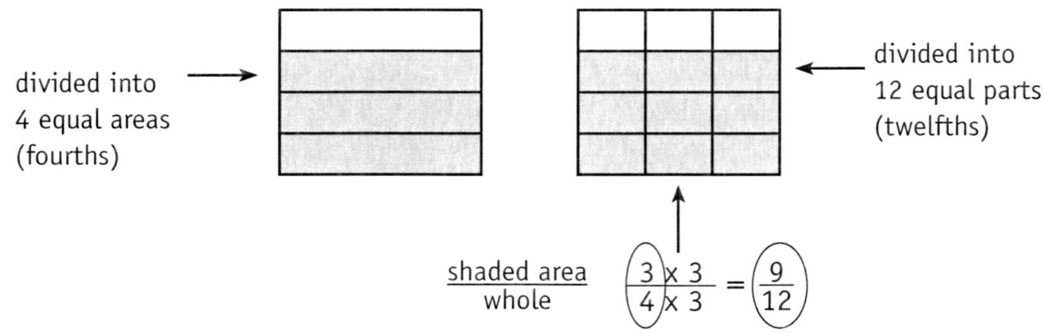

Standard(s)	4.MD.4
Instructional Strategy	Make a line plot.

Students measure their erasers to the nearest $\frac{1}{2}$, $\frac{1}{4}$, or $\frac{1}{8}$ inch. Results are recorded on a line plot.

Number of Erasers Measured

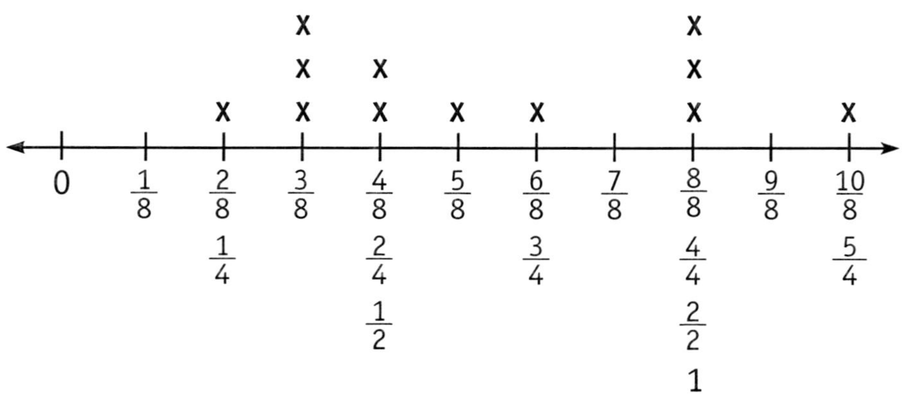

Fifth Grade	
Standard(s)	5.NBT.6, 5.NBT.7
Instructional Strategy	Use concrete models, drawings, and strategies based on place value, the properties of operations, and/or the relationship between multiplication and division. Illustrate and explain the calculation by using equations, rectangular arrays, and/or area models.

Students may use an area model to perform decimal operations.

2.3 x 3.5 = ?

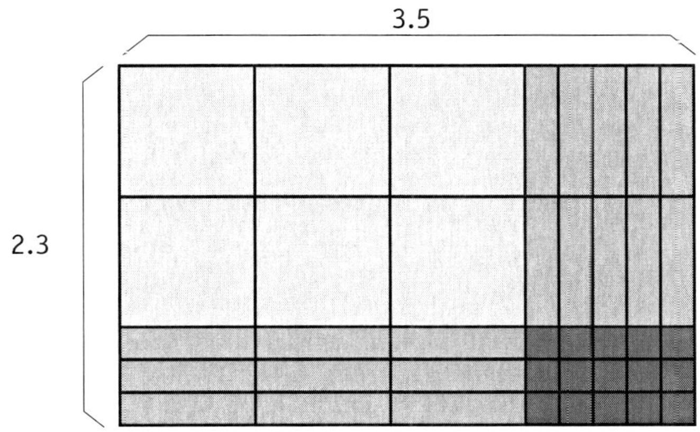

3.5
x 2.3
.15
.90
1.0
+ 6.0
8.05

Standard(s)	5.NF.2, 5.NF.3, 5.NF.4a–b, 5.NF.6, 5.NF.7a–c
Instructional Strategy	Use visual fraction models.

Students may draw a visual fraction model to divide a unit fraction by a whole number.

Three students are working on a math project. One-fourth of the project still needs to be completed. If the work is distributed equally, what fraction of the project does each student have to complete?

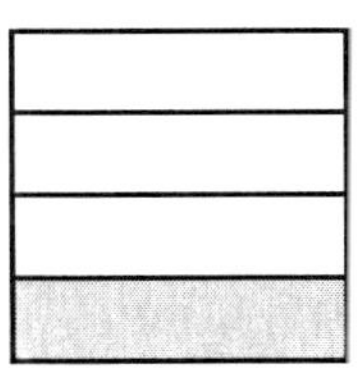

$\frac{1}{4}$ of project

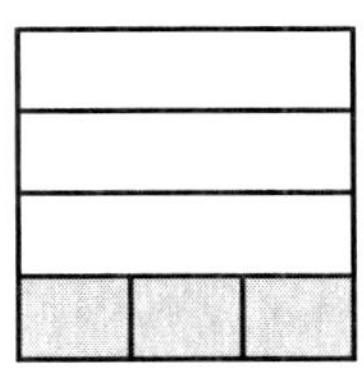

$\frac{1}{4}$ of project split between 3 students

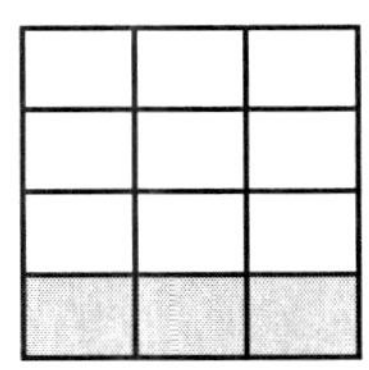

Each student has $\frac{1}{12}$ of the project to complete

$\frac{1}{4} \div 3 = \frac{1}{12}$

$\left(\frac{1}{4} \times \frac{1}{3} = \frac{1}{12}\right)$

Standard(s)	5.MD.5a
Instructional Strategy	Use unit cubes to find volume.

Students may use an empty rectangular prism and fill it with unit cubes to determine its volume.

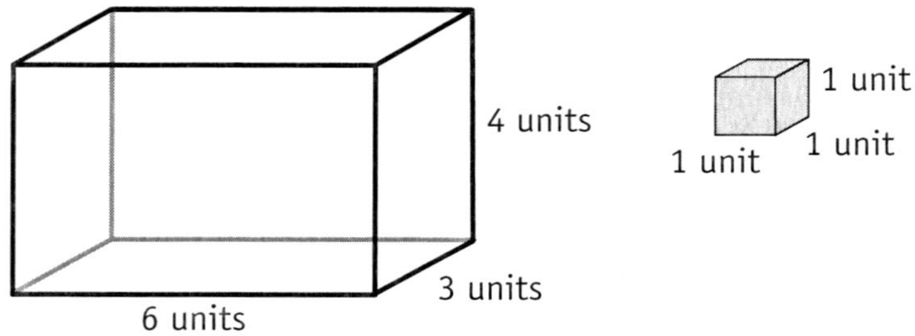

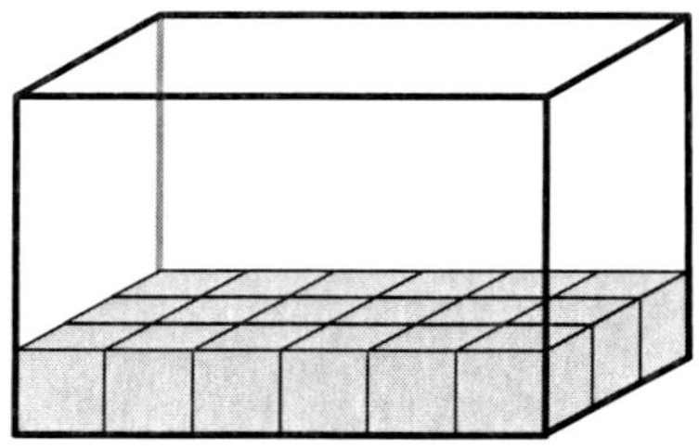

6 x 3 = 18 (1 layer)

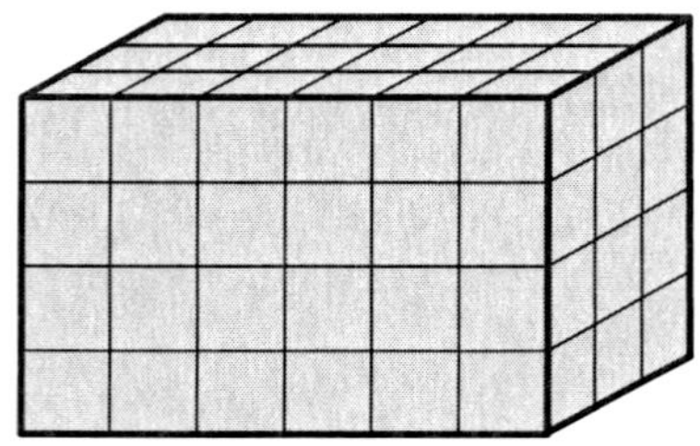

18 x 4 layers = 72 units

Sixth Grade	
Standard(s)	6.RP.3
Instructional Strategy	Reason about tables of equivalent ratios, tape diagrams, double number line diagrams, or equations.

Students may use tape diagrams to illustrate ratios.

A recipe calls for 5 cups of rice for every two cups of water. How many cups of water are needed for 15 cups of rice?

1 piece of tape represents one cup

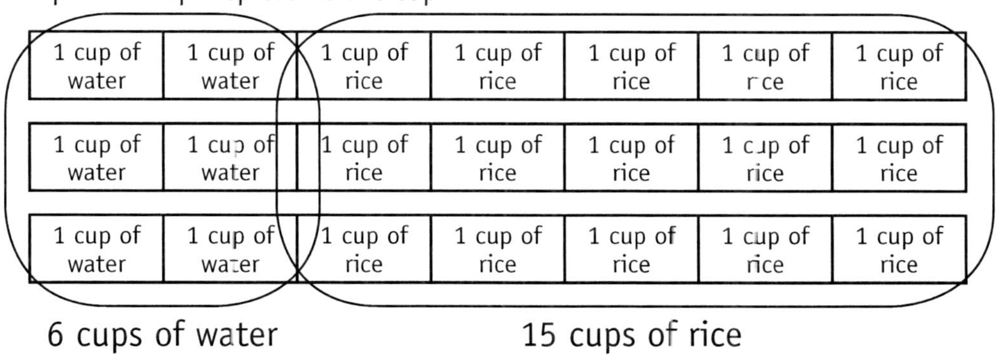

Students may also use a double number line diagram to write ratios.

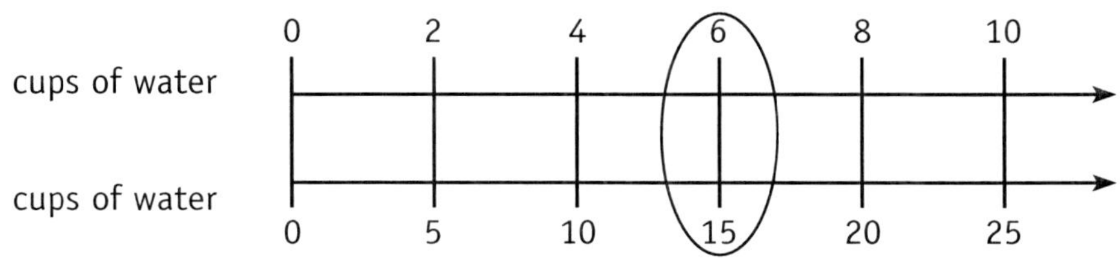

Standard(s)	6.NS.1
Instructional Strategy	Use visual fraction models and equations to represent the problem.

Students may draw a visual fraction model to divide a fraction by a fraction.

Julian has $\frac{3}{4}$ of a yard of fabric to create scarfs that require $\frac{1}{8}$ of a yard of fabric. How many scarfs can he make?

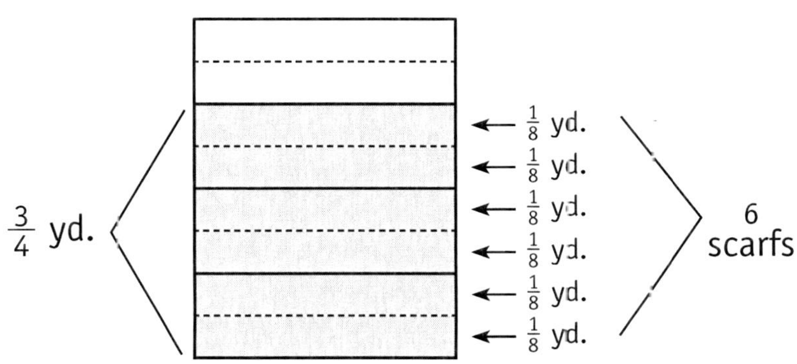

Standard(s)	6.G.2
Instructional Strategy	Find volume by packing with unit cubes.

Students may determine the volume of a rectangular prism by using unit cubes.

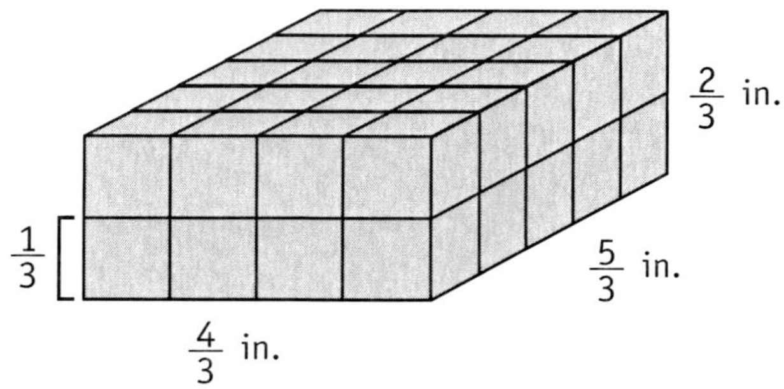

Each cube has a volume of $\frac{1}{3} \times \frac{1}{3} \times \frac{1}{3} = \frac{1}{27}$ in.^{3}

There are 20 cubes on each horizontal layer. There are 40 cube units total.

Volume is $40 \times \frac{1}{27} = \frac{40}{27}$ in.^{3}

Standard(s)	6.SP.4
Instructional Strategy	Plot on a number line, including dot plots, histograms, and box plots.

A box plot shows the descriptive data in an illustrative way. It includes key points such as the median and upper and lower limits. Label the five parts of a box plot. Explain how to find the range of the data.

Box Plot

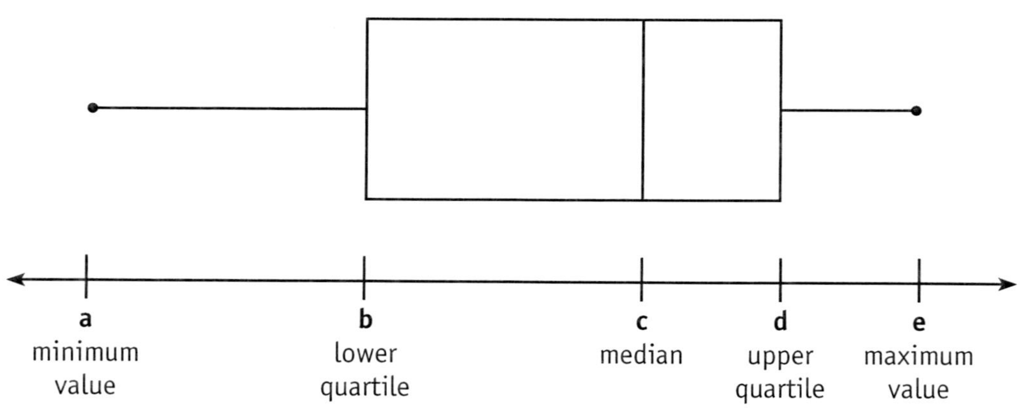

The range is the difference between the maximum value and the minimum value (range = maximum value – minimum value).

Seventh Grade

Standard(s)	7.RP.2
Instructional Strategy	Use tables, graphs, and equations to represent proportional relationships.

Students may draw a graph to decide whether two quantities are in a proportional relationship.

A sandwich calls for 2 slices of bread for every 5 pieces of cold cuts. The graph and equation can be used to represent the proportional relationship between bread and cold cuts.

Cold Cut Sandwiches

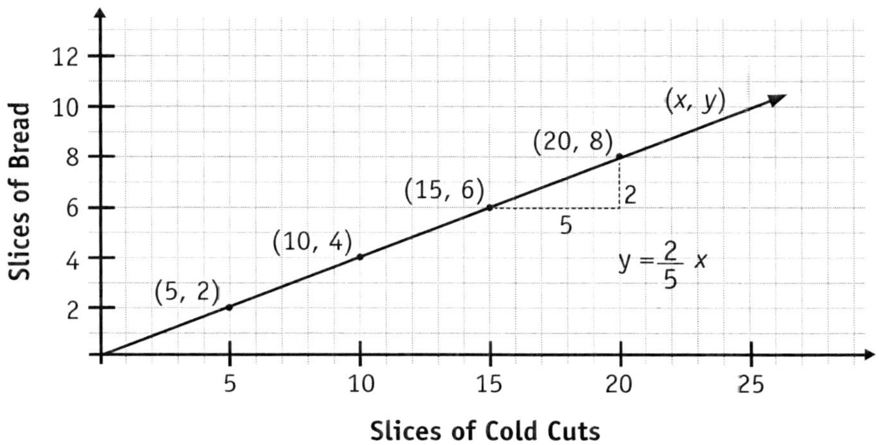

Standard(s)	7.NS.1
Instructional Strategy	Use of number line models or colored chips to understand rational number operations.

Chips of one color can be used to represent positive chips. Chips of another color can be used to represent negative chips.

$-7 + 3 = ?$

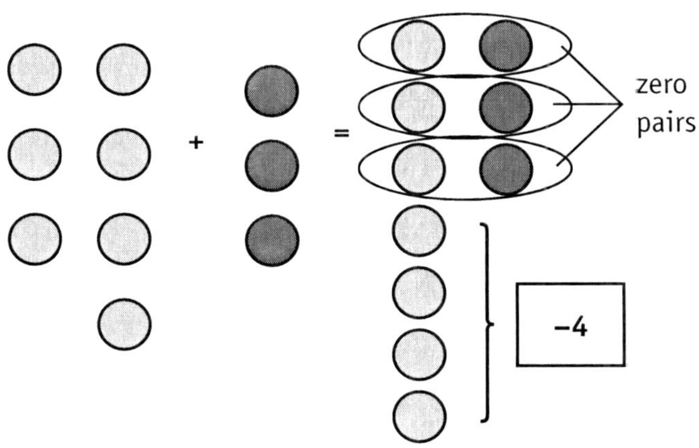

Standard(s)	7.EE.3–4
Instructional Strategy	Use of a tape diagram to solve equations and inequalities.

A tape diagram may be used to represent math situations.

A cell phone bill costs $37.50 every month. There is a monthly flat media fee of $10.50 and a monthly fee for each of the 4 gigabytes of storage on your plan. Write an equation that represents the cost of each gigabyte of storage and find the total cost of the storage.

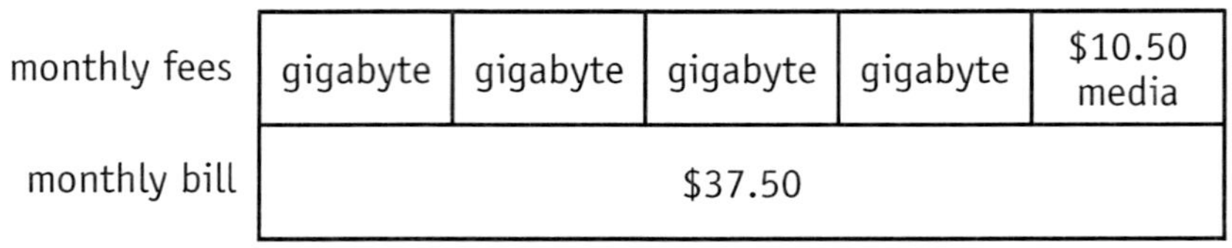

$$4g + 10.50 = 37.50$$

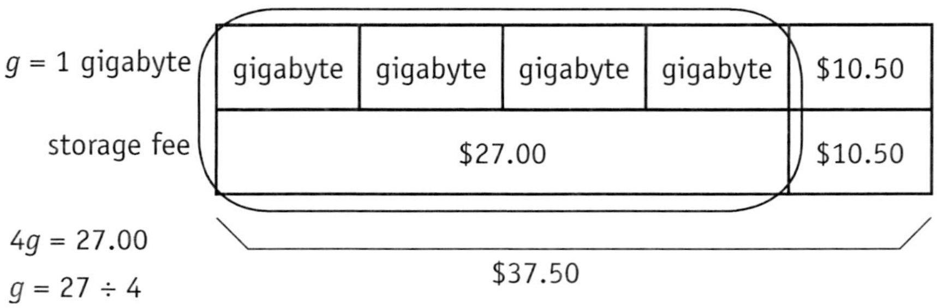

$4g = 27.00$

$g = 27 \div 4$

$g = \$6.75$

Eighth Grade

Standard(s)	8.EE.8
Instructional Strategy	Use drawings to solve systems of linear equations.

Drawings can be used to demonstrate the substitution and/or elimination process for solving systems of equations.

John and his sister Maria have a combined age of 24. John is three times as old as Maria. What are their ages?

4 Maria's = 24, therefore Maria is 6 years old. John is 3 times as much, or 18 years old.

Standard(s)	8.G.1–3
Instructional Strategy	Use of models, transparencies, geo-boards, and software to understand transformations.

Students determine if shapes are congruent by moving physical objects (models, geo-boards, software) using reflections, rotations, and/or translations.

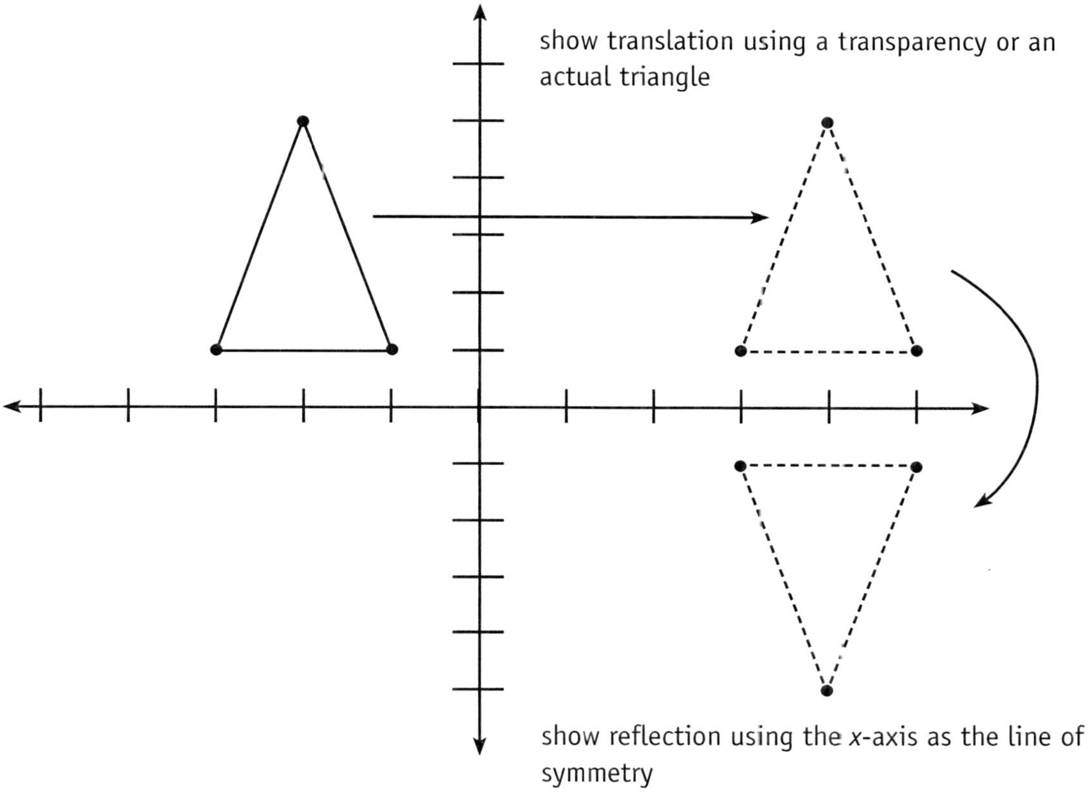

Standard(s)	8.SP.4
Instructional Strategy	Display frequencies and relative frequencies in a two-way table.

A two-way table is a table that shows the frequency of two variables. Rows indicate one variable. Columns indicate a different variable.

Forty high school students were asked about which Los Angeles school they prefer.

Gender	USC	UCLA	Neither
Male	12	5	1
Female	11	6	5

It appears that most students, regardless of gender, prefer USC.

Let us now explore practical applications of a few of these instructional strategies in combination with a selected math task.

Example Grades K–1

Application Task: Jada had 12 crayons. She gave 5 to her sister. How many crayons does she have now?

Selected Strategy: Decomposing a number leading to ten.

Rationale: A number greater than 10 is given that is being reduced. The relationship between this number, 12, and 10 (2 units) can be used to show how many parts are left from 5 when the 2 units are deducted.

Student Application:

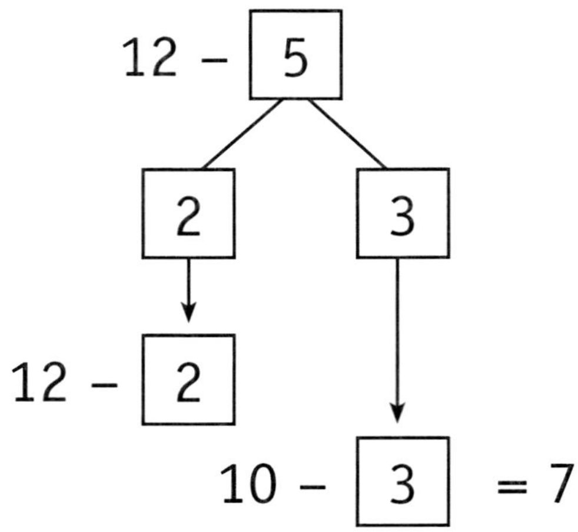

Example Grades 2–3

Application Task: Maria ate $\frac{5}{8}$ of a blueberry pie. Susan ate $\frac{2}{4}$ of a blueberry pie that is the same size. Who ate more pie? Explain.

Selected Strategy: Draw a fraction area model.

Rationale: Using the same fraction model to represent both fractions may help illustrate which area is larger.

Student Application:

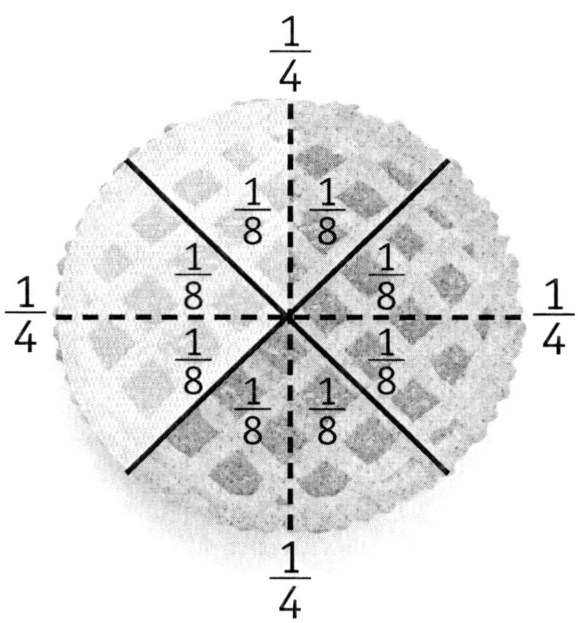

$\frac{5}{8}$ is larger than $\frac{2}{4}$

5 one-eighths shaded is more than 2 one-fourths shaded

Example Grades 4–5

Application Task: A recipe for 3 batches of donuts calls for $\frac{1}{2}$ tablespoon of vanilla. Mike only wants to make one batch of donuts. How many tablespoons of vanilla should he use? Explain your thinking.

Selected Strategy: Use of a fraction model.

Rationale: Shows a visual representation of a part of a whole being partitioned even further. The model will show the relationship between this new part and the original whole.

Student Application:

$$\frac{1}{2} \div 3 = ?$$

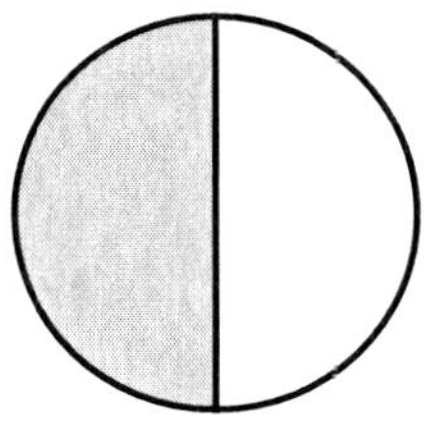

$\frac{1}{2}$ tablespoon

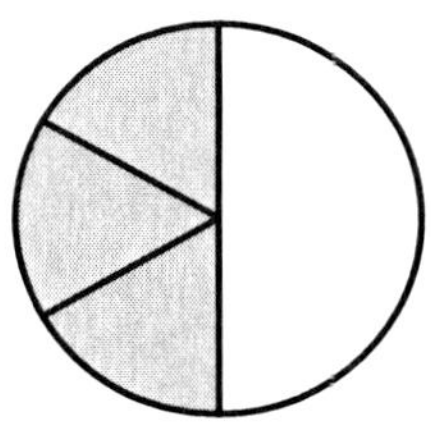

$\frac{1}{2} \div 3$

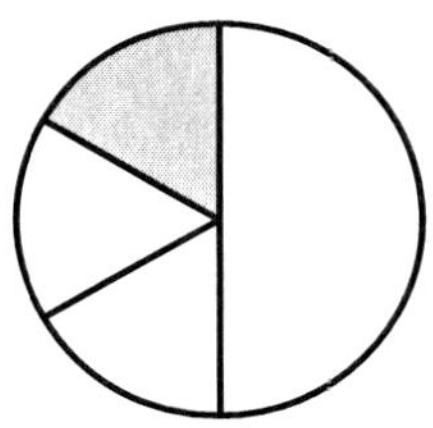

only one is needed

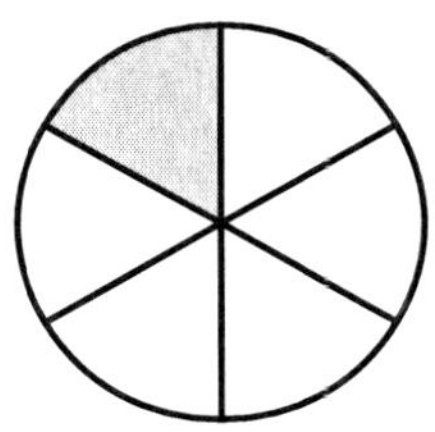

$\frac{1}{6}$ of a tablespoon is needed

One-half of the circle is shaded. I split the $\frac{1}{2}$ into 3 parts. I only needed one of the three parts. I thought, "How much of the entire circle does this 1 part mean?" It is $\frac{1}{6}$ of the whole circle.

Example Grades 6–8

Application Task: Jessica's briefcase weighs 26 lbs. She has 3 algebra books (all of which are the same weight) and one laptop in the briefcase. The laptop weighs 5 lbs. How much does each algebra book weigh?

Selected Strategy: Use a tape diagram.

Rationale: The tape diagram will help the student visualize the relationship between the quantities.

Student Application:

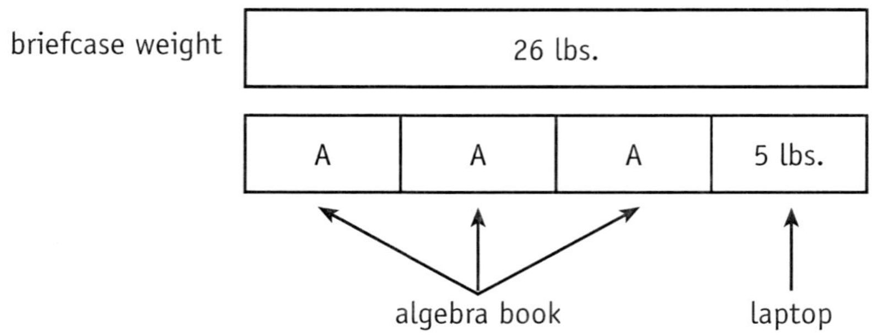

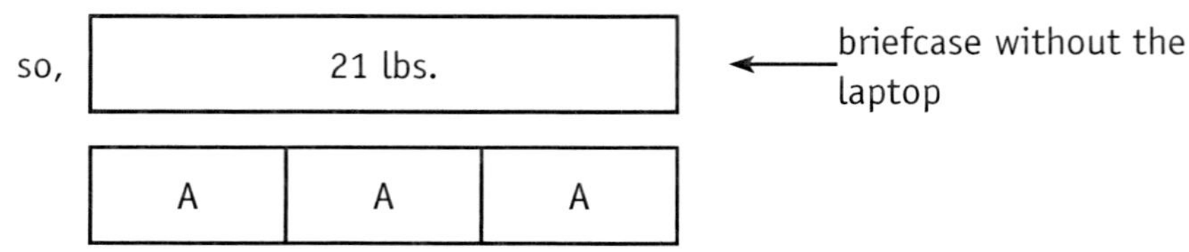

I know that 7 + 7 + 7 = 21, so each algebra book weighs 7 lbs.

Mathematical Strategies Protocol

Think about which instructional strategies and models presented in Figure 5.6 are applicable to your grade level. Follow these easy steps to select mathematics models for a selected task. This activity can be done independently or in grade-level teams.

Step 1: Choose a mathematical concept from your grade-level standards.

Step 2: Write or select a task that students will solve.

Step 3: Decide which instructional strategies and/or models you might encourage students to use to solve the task.

Step 4: Describe your rationale for choosing those strategies and/or models.

Step 5: Provide a sample that shows how a student would apply those strategies and/or models.

You Try It!

1. Download the Mathematics Strategies template from the Digital Resource CD (filename: strategies.pdf)
2. Review the sample mathematical strategies for your grade level (page 166–185).
3. Follow the protocol to complete the chart.
4. Collaborate with your grade-level colleagues to share your ideas and consider additional mathematical models for your selected task.

Unit Checklist

Now that you have learned specific methods for analyzing and developing units of study, it is important to have a checklist to use as a reminder of all of the essential pieces that make up a successful unit. This unit checklist can be used in grade-level teams to analyze existing units of study or units that you create. The questions assess the use of the instructional strategies of focus, coherence, and rigor. An analysis of grade-level content and strategic use of the mathematical practices is also included.

My recommendation is to analyze a unit prior to implementation to make proper adjustments, if needed. This tool can also be used as a point of reflection after instruction.

Figure 5.7 Unit Checklist

- ☐ Is the content aligned with grade-level expectations—Focus?
- ☐ Is there an explicit connection between the MPs and the content?
- ☐ Are the MPs explicitly applied in the unit of study?
- ☐ Is there a balance between the MPs represented in the unit?
- ☐ Is the entire curriculum weighted according to the shift of focus?
- ☐ Are lessons targeting additional and supporting clusters linked to major clusters?
- ☐ Are the individual lessons tied to the clusters (e.g., learning objectives reinforced throughout the lesson)?
- ☐ Are the lessons designed so that prior knowledge becomes connected to the new knowledge?
- ☐ Does an equal balance exist among problem types (e.g., procedural, conceptual, and application)?
- ☐ Are algorithms based on mathematical foundations and not mnemonic devices?
- ☐ Are there multiple lesson types (e.g., direct instruction, inquiry-based) used in the unit?
- ☐ Is there an explicit connection between the selected model of teaching and the skill/concept being targeted?

Stop and Think

Question 1

When might a teacher incorporate a direct instruction model versus an inquiry-based investigative model in the unit?

Question 2

How might the teacher provide the time for students to investigate new concepts in an inquiry-based model? How does the instructional shift of focus provide relief regarding decisions of time?

Question 3

How might you incorporate additional models of teaching beyond a direct instruction and inquiry-based investigative model?

Chapter 6

Assessment

Assessment is used to monitor students' progress toward established learning objectives and to create new learning experiences as a result of the outcome. With the advent of today's standards, authentic assessments have been created to capture meaningful learning of mathematics. Assessment is aligned with the learning experience, which in turn is aligned with the learning objective/standards, thus redefining the notion of "teaching to the test." The two national consortia charged with creating these authentic assessments are the Partnership for Assessment of Readiness for College and Careers (PARCC) and the Smarter Balanced Assessment Consortium (SBAC). Both consortia offer a long-range end-of-the-year summative assessment which captures a student's cumulative grade-level knowledge and interim benchmark formative assessments which serve as progress indicators regarding specific content. Each consortium works closely with the state education chiefs and officials to ensure that the assessment system meets the needs of the participating states and the demands of the standards. State assessment affiliation can be found by visiting your respective department of education website. This chapter focuses on assessment types that are aligned to the standards.

The following resources are included in this chapter to not only help your students prepare for the state assessment, but also develop as mathematicians.

- **Assessment Question Types**—A range of question types are used to measure meaningful learning of mathematics. This section outlines the various question types and provides tools to embed these measures into existing curricula.
- **Curriculum Assessment Analysis**—This curriculum assessment analysis allows for an inventory of existing curricula. Measures can be taken to reinforce identified deficits to align with the next generation assessment characteristics.
- **Procedural Fluency Assessment**—Procedural fluency is one of the three components of rigor. The standards specifically describe where fluency is expected. This section provides grade-level fluency assessment to complement concept and application development.

Assessment Question Types

The next generation assessments, such as PARCC and SBAC, offer a range of question types to measure meaningful learning (not just rote learning) of both the content standards and standards for mathematical practice. Many of the assessments make use of a technology-based format.

The next generation assessments use a range of question types, including selected response (SR), constructed response (CR), extended construction response (ER), technology enhanced (TE), and performance task (PT). These questions are designed to measure a deeper understanding and application of mathematical concepts. A description of each question type, examples, and curriculum adaption strategies are outlined as follows.

Selected Response (SR)

Selected-response questions measure a student's understanding of one or two mathematical concepts. Students are asked to select one or more responses from a set of multiple-choice options. Not all selected-response questions will have more than one correct response. Students will not necessarily be prompted to know how many correct responses are present. Classic multiple-choice questions are also a part of the selected-response question types. Here are a few SR question types.

Grade 3

For numbers 1a to 1d, choose Yes or No to indicate whether each number graphed on the number line represents one whole.

1a. 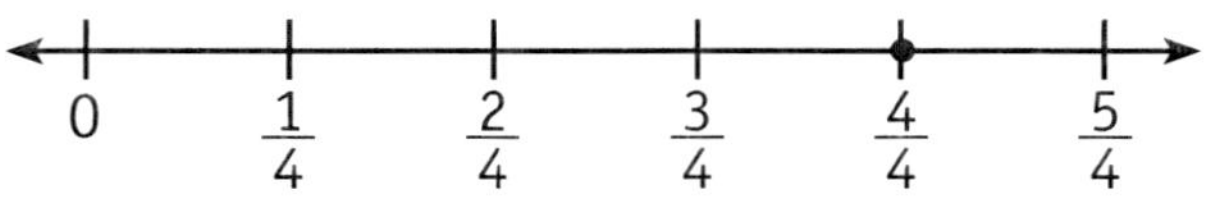

○ Yes ○ No

1b. 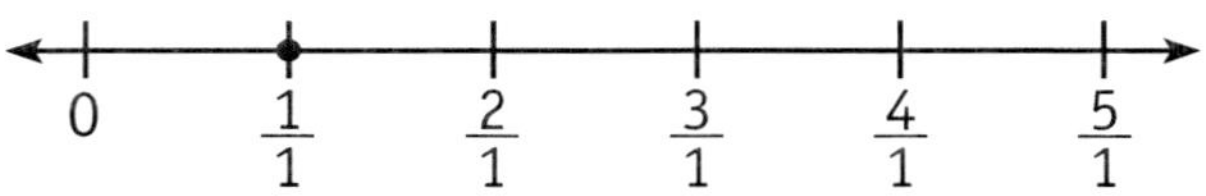

○ Yes ○ No

1c. 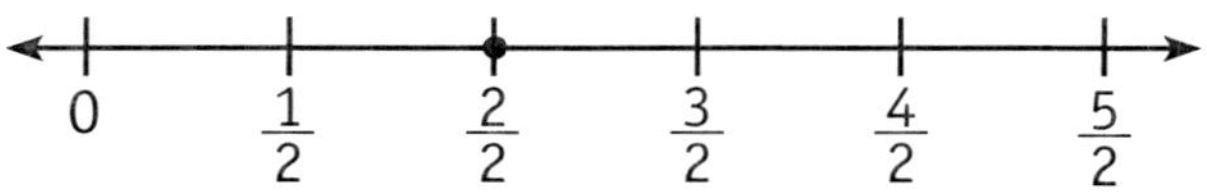

○ Yes ○ No

1d. 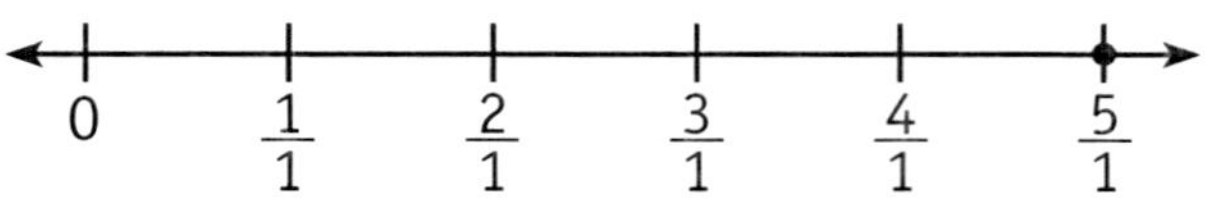

○ Yes ○ No

Grade 3

For numbers 1a–1d, state whether each figure has $\frac{3}{5}$ of its whole shaded.

1a. 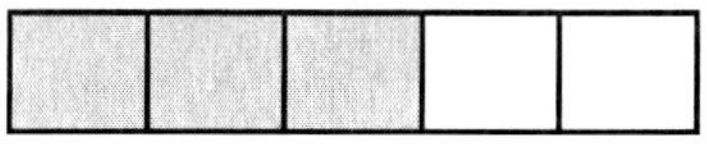 Ⓨ Yes Ⓝ No

1b. 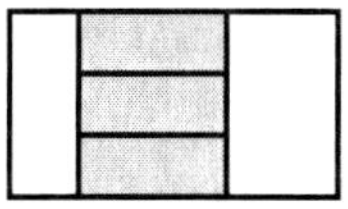 Ⓨ Yes Ⓝ No

1c. 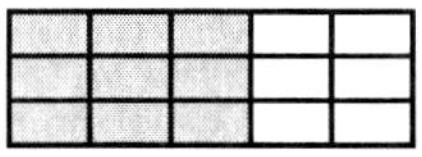 Ⓨ Yes Ⓝ No

1d.  Ⓨ Yes Ⓝ No

Grade 5

On Saturday Conni ran $3\frac{6}{10}$ miles. On Sunday she ran $1\frac{1}{2}$ miles. How many miles did Conni run all together?

For numbers 1a–1d, select Yes or No to indicate whether each equation can be used to solve the word problem shown above.

1a. $3\frac{6}{10} + 1\frac{5}{10} = \square$ Ⓨ Yes Ⓝ No

1b. $\frac{30}{10} + \frac{10}{10} = \square$ Ⓨ Yes Ⓝ No

1c. $\frac{36}{10} + \frac{15}{10} = \square$ Ⓨ Yes Ⓝ No

1d. $3\frac{6}{10} + 1\frac{1}{10} = \square$ Ⓨ Yes Ⓝ No

Grade 7

Raj made a graph that represents the amount of money he earns, y, for the numbers of hours he works, x. The graph is a straight line that passes through the origin and the point (1, 10.5).

Which statement **must** be true?

A. The y-intercept of the graph is 10.5.

B. Raj earns $10.50 per hour.

C. The slope of the graph is 1.

D. Raj works 10.5 hours per day.

Notice that not all SR questions have more than one correct response. Your existing curriculum may not provide SR multiple-choice questions in which there is more than one correct response. Modification may be needed in the absence of supplemental resources. The following is an example of an existing problem that has been modified to create more than one possible answer choice.

Existing problem:

Grade 5

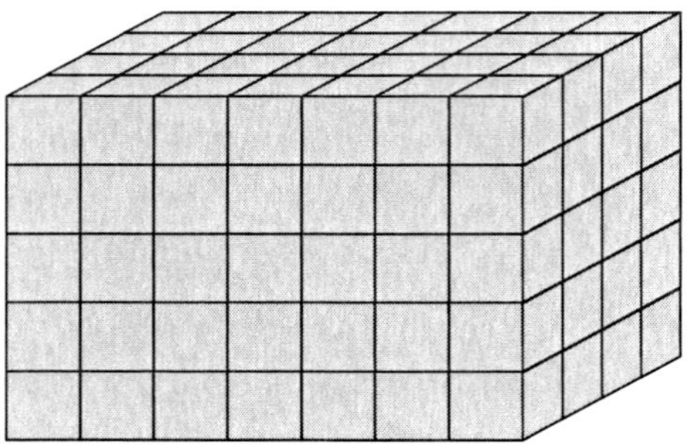

Amelia counted the blocks to find the volume of the figure above. Which of the following might be another way for her to find the volume?

A. Add 7 + 4 + 5

B. Multiply 7 × 5 × 5

C. Multiply (7 × 4) × 5

D. Add 7 + 5 + 4 and multiply by 2

Notice that the problem provides one correct answer. The words "another way" alert the teacher that this can be modified. Multiple steps can be taken that lead to the correct solution. Notice the change that has been made.

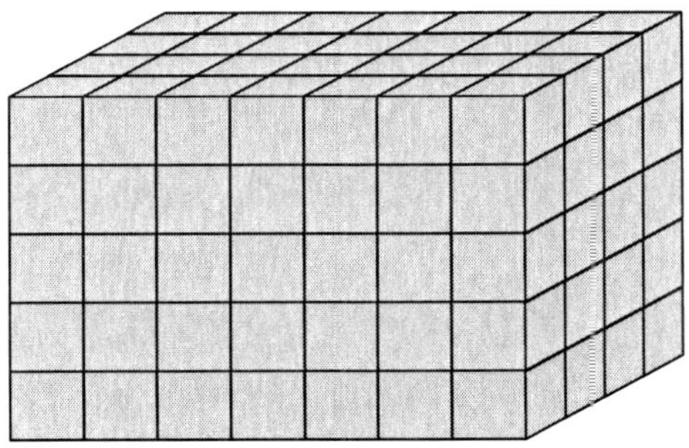

Amelia counted the blocks to find the volume of the figure above. Which of the following might be another way for her to find the volume?

A. Add 7 + 4 + 5

B. Multiply 7 × 5 × 5

C. Multiply (7 × 4) × 5

D. Add 7 + 5 + 4 and multiply by 2

E. Count the blocks in the top layer and multiply by 5

An additional response has been included that assesses the student's depth of knowledge regarding this concept. For scaffolding purposes, you may want to insert language into the prompt that indicates the possibility of there being more than one correct answer. You may choose to insert "choose all that apply." This scaffold, if inserted, should eventually be removed over time.

It is important to note that not all multiple-choice questions must have more than one correct answer. As you are modifying your curriculum, look for natural fits where changes can be made. It may not be advantageous to change certain problems based on the structure of the question. This activity can be performed in grade-level teams and serves to strengthen your content knowledge and the ability to design curricula.

Constructed Response (CR)

Constructed-response questions measure a student's understanding of one or two mathematical concepts similar to a selected response question. However, students do not have options from which to choose. They must "construct" their response by providing a text or numerical solution. A multiple-choice question can be modified to create a CR question type by omitting the answer choices. The following are examples of CR questions.

Grade 3

Alyssa is making a quilt for her brother. The shaded parts are complete.

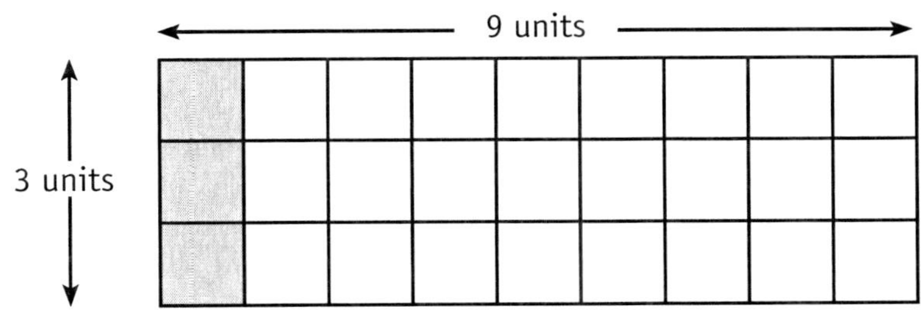

What will the area be when the whole quilt is complete?

Grade 5

Coach Hamm calculated the total number of miles the team ran during September, October, and November.

Month	Number of Miles
September	62.25
October	72.5
November	76.75

A. What is 10 times the number of miles the team ran in September?

B. What is $\frac{1}{10}$ of the number of miles the team ran in October?

Grade 8

Once the summer season is over, the Belleview City Works Department is responsible for draining and cleaning all of the city pools. There are 10 pools in the city and each pool holds 24,000 gallons of water. Each hour, 400 gallons of water can be drained.

Write an equation that models how long it takes to drain a pool in the city of Belleview. Let h = the number of hours and g = the number of gallons remaining in the pool.

Extended Response or Extended Constructed Response (ER)

Extended constructed response questions measure a student's understanding of one or two mathematical concepts similar to a constructed-response question. However, students must produce a more extended response that includes an explanation of how the solution was derived with a viable argument. Rubrics are often used in scoring.

Modifications to your existing curriculum may include building upon existing constructed-response questions. This can be achieved by prompting students to defend and justify their answers. Teachers can also create an extension question by asking a second or third question that is related to or dependent upon information from the first question.

The following are examples of ER questions.

Grade 3

There will be a pancake breakfast at Park View School. Eight tables are available. Each table can seat 6 people. How many people can be seated at the breakfast? How do you know this?

Solution:

Explain your solution:

Chapter 6

Grade 5

Ms. Tan has to rearrange bookshelves to make room for the canned food drive. Ms. Tan needs to know how many books are on the shelves. Cory and Albert volunteer to count the books in a double-sided bookcase and make a chart.

Shelf Number	Books (Side 1)	Books (Side 2)
1	11	17
2	11	17
3	11	17
4	11	17
5	11	17

A. Cory says that she knows there are 140 books on the first five shelves of the bookcase. She explains, "I added 11 and 17, and then I multiplied by 5." Is Cory correct? Explain your thinking.

B. Albert says, "I got the same number. First, I wrote the number sentence $5 \times (11 + 17)$, and then I found the solution." Is Albert correct? How do you know?

Grade 6

Mr. O'Donnell told his class that they will be buying some plants for their classroom. The cost of the plants and a watering can is represented by the expression $4p + 12$.

A. Identify the variable in this expression. What does it represent?

B. Identify the coefficient. What operation does the expression $4p$ tell you to perform?

Mr. O'Donnell tells the students that they will also need to pay sales tax, which is 8%.

Write an expression that represents the amount of sales tax they will pay. Explain your answer.

Technology-Enhanced/Technology-Embedded (TE)

Technology-enhanced questions not only assess a student's knowledge of the content and the mathematical practices, but also their ability to navigate the technology platform. This includes exercises such as maximizing/minimizing, clicking and dragging, and constructing shapes. These question types assess a student's ability to manipulate multimedia and interactive tools. Here are a few examples of TE question types.

Grade 3

The picture shows Ethan's flower garden. Fill in the box to make a fraction that represents the part of Ethan's garden that is covered with flowers.

$\frac{6}{\square}$ Which letter represents this fraction's location on the number line?

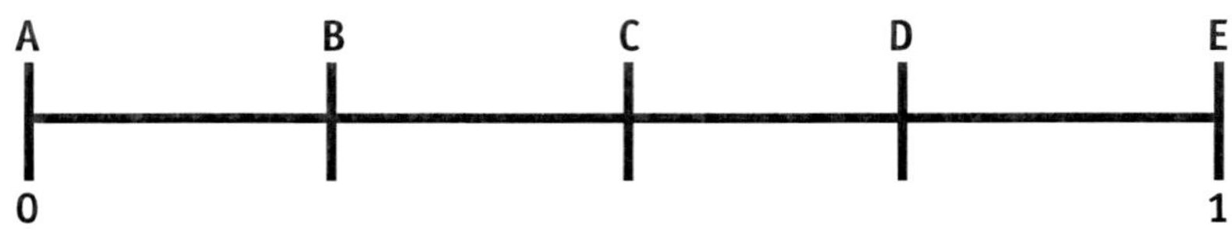

Your answer: ☐

Drag tiles to build fractions that make the statement true. There is more than one correct answer.

4	6	8

$\frac{3}{\square} < \frac{3}{\square}$

Grade 5

Mary is installing ceramic art squares on a rectangular canvas.

- She is using ceramic squares.
- The length of a side of each ceramic square is $\frac{1}{2}$ foot.
- The area of the canvas is 26 square feet.
- The width of the canvas is 4 feet.

Use the grid to model the rectangular canvas that Mary is installing.

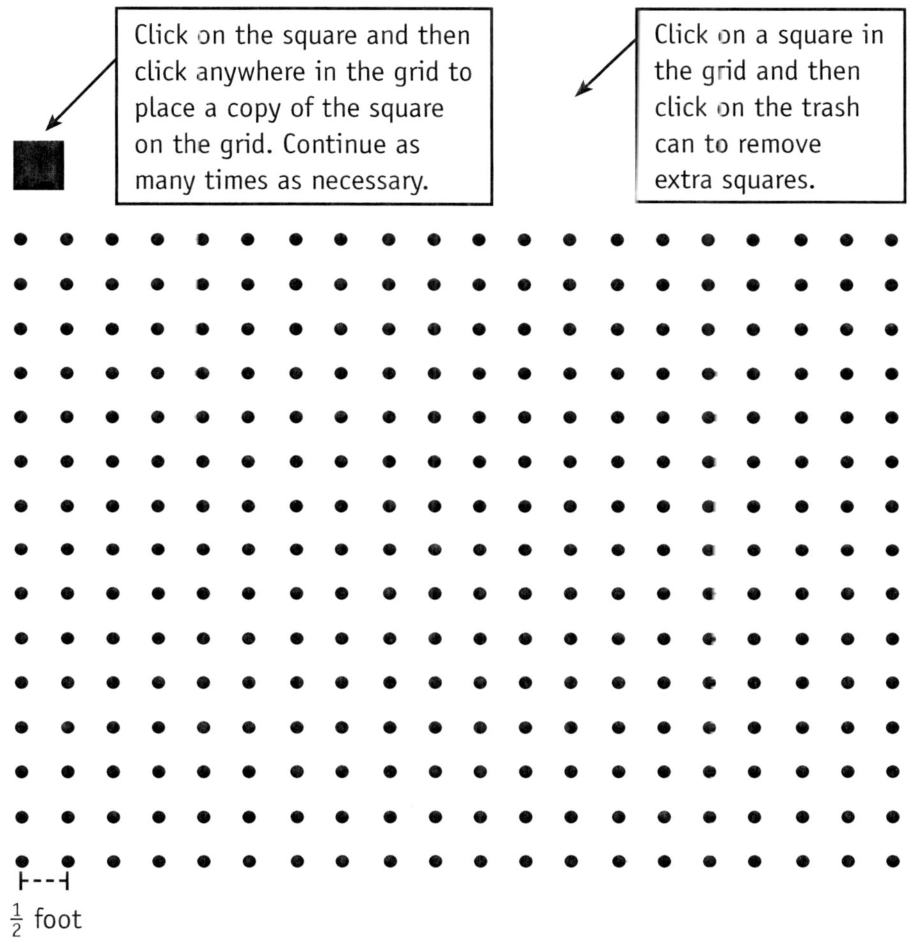

What is the length, in feet, of the floor? ☐

Grade 8

Drag and construct line *b* on the coordinate grid so that the following is true:

- line *a* and line *b* represent a system of linear equations with a solution of (6, –3).
- the slope of line *b* is greater than –1 and less than 1.
- the *y*-intercept of line *b* is positive.

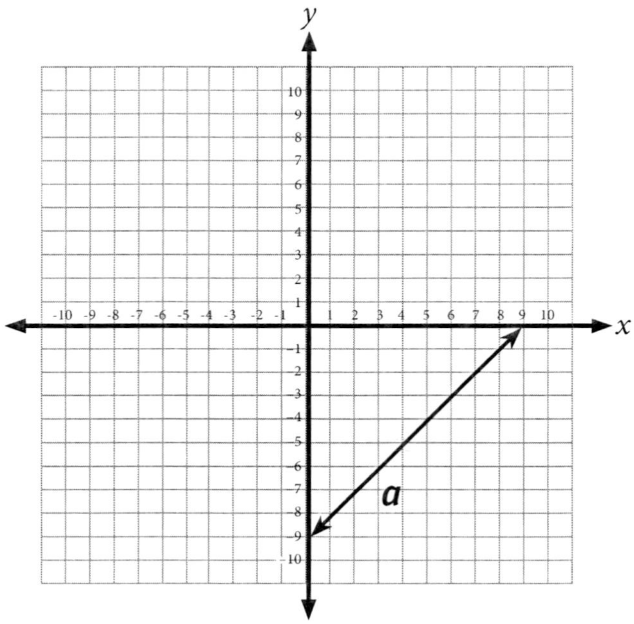

Students must have exposure to technology in order to become familiar with the skills needed to navigate the assessment platform. Students may not be able to demonstrate what they know if they are not accustomed to manipulating the technology. Students should spend time using computer keyboards to input text and data, maximizing and minimizing objects on a screen, and using calculators embedded for specific problems. The standard for Mathematical Practice 5 requires strategic tool selection. Embedding technology in curriculum and instruction will enable students to develop this skill.

Performance Task (PT)

Performance tasks assess multiple concepts across different mathematics clusters and domains. Performance tasks include multiple parts that work in an integrated fashion. The domino effect is often in place: successful progression through subsequent portions of the performance task is dependent upon previous answers derived throughout the task. Performance tasks are bound in real-world contexts and primarily assess two dimensions of rigor (conceptual knowledge and application). Time allotment for a task ranges from 80 minutes to 120 minutes based on the grade level and task. The following highlights features of a performance task.

A performance task assesses knowledge across multiple content domains, includes multiple standards for mathematical practice, and has questions that are interdependent.

Grade 4 Example

Title	Playing Around
Grade	4
Content Domain(s)	Operations and Algebraic Thinking Number Operations–Fractions
Standards for Mathematical Practice	MP1 Make sense of problems and persevere in solving them. MP3 Construct viable arguments and critique the reasoning of others. MP7 Look for and make use of structure.
Assessment Target(s)	• Construct, autonomously, chains of reasoning that will justify or refute propositions or conjectures. • Identify important quantities in a practical situation and map their relationships (e.g., using diagrams, two-way tables, graphs, flowcharts, or formulas). • Select and use appropriate tools strategically. • Apply mathematics to solve well-posed problems arising in everyday life, society, and the workplace.
Score	Use Rubric Scoring Guide
Purpose	Assess a student's ability to problem-solve with fractions.

Playing Around

Part A

Mrs. Shannon's class is performing a class play.

1. Mikey and Lexi both practice their lines for the play. Mikey practices his lines for $\frac{3}{4}$ of an hour every night. Lexi practices for $\frac{1}{2}$ of an hour daily. Who practices their lines for a longer time? Use a fraction model to show your thinking.

Solution: ______________________________

2. Brie and Doug are helping to set up the chairs in the auditorium. Mrs. Shannon says that the chairs need to have $\frac{2}{3}$ of a foot between them. Brie and Doug want to know the equivalent fraction in twelfths, so they will know how many inches this is equal to. What is the equivalent fraction in twelfths? Use a fraction model to show your thinking.

Solution: ______________________________

Chapter 6

Part B

1. Kenzie, Sherry, and Juan are painting props for the play. Each prop uses $\frac{1}{3}$ of a can of paint. If they need to paint 8 props, how many cans of paint do they need?

Solution: ____________________

2. Matt and Omar are making a poster to tell the school about the play. They start with a sheet of butcher paper that is $\frac{11}{12}$ of a meter long. They cut off a piece that is $\frac{6}{12}$ of a meter. What fraction of the whole paper is left?

Solution: ____________________

Part C

1. The class sold tickets to the show this week. There are 100 seats in the auditorium. The chart shows the number of seats sold each day. What fraction of the total number of seats is sold?

Monday	Tuesday	Wednesday	Thursday
21 seats	19 seats	27 seats	20 seats

Solution: ______________________________

Rewrite the fraction as a decimal. Use the hundredths grid as a model.

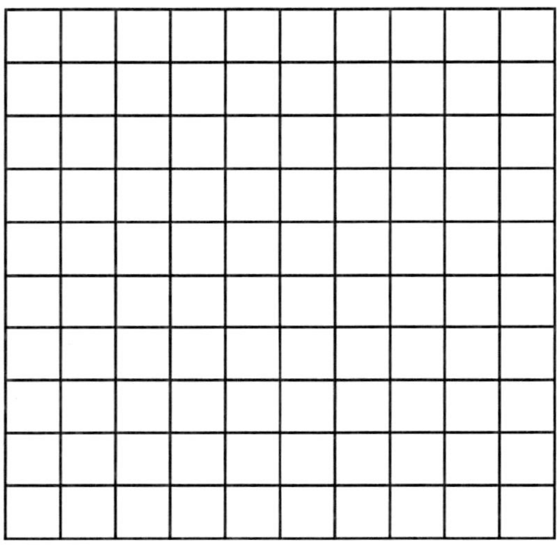

Solution: ______________________________

Creating Assessment Tasks

Assessment tasks should not just focus on action verbs noting cognitive demand but also the mathematical context that supports the cognitive exploration. The Hess Cognitive Rigor Matrix (Figure 6.1) shows the intersection of Bloom's Taxonomy and Webb's Depth of Knowledge levels (Hess et. al 2009). The table provides specific mathematical tasks aligned to each DOK level and can help you craft your own authentic assessments based on the suggestions offered. It is important to note that students should have the opportunity to explore tasks within the entire range of DOK levels. Noting deficits in students' ability to negotiate different DOK tasks should determine levels to target during curriculum selection or design.

Figure 6.1 Hess' Cognitive Rigor Matrix

Revised Bloom's Taxonomy	Webb's DOK Level 1 Recall and Reproduction	Webb's DOK Level 2 Skills and Concepts	Webb's DOK Level 3 Strategic Thinking/ Reasoning	Webb's DOK Level 4 Extended Thinking
Remember Retrieve knowledge from long-term memory, recognize, recall, locate, identify	Recall, observe, and recognize facts, principles, and properties Recall/identify conversions among representations or numbers (e.g., customary and metric measures)			
Understand Construct meaning, clarify, paraphrase, represent, translate, illustrate, give examples, classify, categorize, summarize, generalize, infer a logical conclusion, predict, compare/contrast, match like ideas, explain, construct models	Evaluate an expression Locate points on a grid or number on a number line Solve a one-step problem Represent math relationships in words, pictures, or symbols Read, write, and compare decimals in scientific notation	Specify and explain relationships (e.g., non-examples/examples; cause-effect) Make and record observations Explain steps followed Summarize results or concepts Make basic inferences or logical predictions from data/observations Use models/diagrams to represent or explain mathematical concepts Make and explain estimates	Use concepts to solve <u>non-routine</u> problems Explain, generalize, or connect ideas <u>using supporting evidence</u> Make <u>and justify</u> conjectures Explain thinking when more than one response is possible Explain phenomena in terms of concepts	Relate mathematical or scientific concepts to other content areas, other domains, or other concepts Develop generalizations of the results obtained and the strategies used (from investigation or readings) and apply them to new problem situations

Figure 6.1 Hess' Cognitive Rigor Matrix (cont.)

Revised Bloom's Taxonomy	Webb's DOK Level 1 Recall and Reproduction	Webb's DOK Level 2 Skills and Concepts	Webb's DOK Level 3 Strategic Thinking/ Reasoning	Webb's DOK Level 4 Extended Thinking
Apply Carry out or use a procedure in a given situation; carry out (apply to a familiar task), or use (apply) to an unfamiliar task	Follow simple procedures (recipe-type directions) Calculate, measure, apply a rule (e.g., rounding) Apply an algorithm or formula (e.g., area, perimeter) Solve linear equations Make conversions among representations or numbers, or within and between customary and metric measures	Select a procedure according to criteria and perform it Solve routine problems, applying multiple concepts or decision points Retrieve information from a table, graph, or figure and use it to solve a problem requiring multiple steps Translate between tables, graphs, words, and symbolic notations (e.g., graph data from a table) Construct models given criteria	Design investigation for a specific purpose or research question Conduct a designed investigation Use concepts to solve non-routine problems <u>Use and show reasoning, planning, and evidence</u> Translate between problem and symbolic notation when not a direct translation	Select or devise approach among many alternatives to solve a problem Conduct a project that specifies a problem, identifies solution paths, solves the problem, and reports results
Analyze Break into constituent parts, determine how parts relate, differentiate between relevant-irrelevant, distinguish, focus, select, organize, outline, find coherence, deconstruct	Retrieve information from a table or graph to answer a question Identify whether specific information is contained in graphic representations (e.g., table, graph, T-chart, diagram) Identify a pattern/trend	Categorize and classify materials, data, and figures based on characteristics Organize or order data Compare/contrast figures or data Select appropriate graph and organize and display data Interpret data from a simple graph Extend a pattern	Compare information within or across data sets or texts Analyze and <u>draw conclusions from data, citing evidence</u> Generalize a pattern Interpret data from a complex graph Analyze similarities/ differences between procedures or solutions	Analyze multiple sources of evidence Analyze complex/abstract themes Gather, analyze, and evaluate information

Figure 6.1 Hess' Cognitive Rigor Matrix (cont.)

Revised Bloom's Taxonomy	Webb's DOK Level 1 Recall and Reproduction	Webb's DOK Level 2 Skills and Concepts	Webb's DOK Level 3 Strategic Thinking/ Reasoning	Webb's DOK Level 4 Extended Thinking
Evaluate Make judgments based on criteria, check, detect inconsistencies or fallacies, judge, critique			Cite evidence and develop a logical argument for concepts or solutions Describe, compare, and contrast solution methods Verify reasonableness of results	Gather, analyze, and evaluate information to draw conclusions Apply understanding in a novel way, provide argument or justification for the application
Create Reorganize elements into new patterns/structures, generate, hypothesize, design, plan, produce	Brainstorm ideas, concepts, or perspectives related to a topic	Generate conjectures or hypotheses based on observations or prior knowledge and experience	Synthesize information within one data set, source, or text Formulate an original problem given a situation Develop a scientific/ mathematical model for a complex situation	Synthesize information across multiple sources or texts Design a mathematical model to inform and solve a practical or abstract situation

(Hess 2009)

Curriculum Assessment Analysis

Your existing assessments should be aligned to the design approaches presented in the next generation assessments. An audit of your existing assessments is needed in order to reinforce possible deficits and strengthen the tools used to capture students' understanding of mathematics.

The Curriculum Assessment Analysis template located on the Digital Resource CD can be used to analyze existing resources to determine alignment with the next-generation assessment system. The analysis can be completed in grade-level teams to determine what additional resources are needed to reinforce identified assessment deficits.

Figure 6.2 Curriculum Assessment Analysis Template

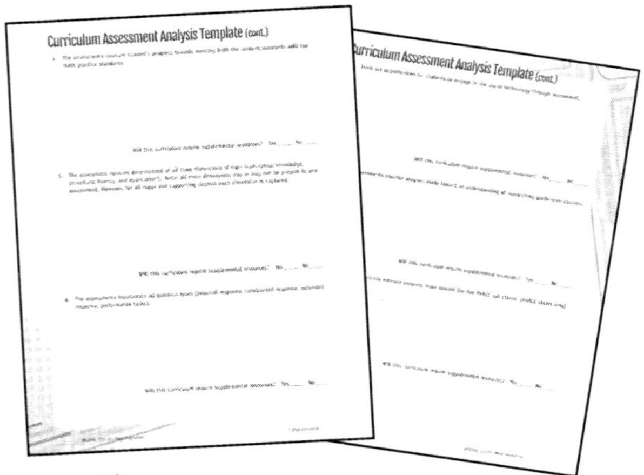

1. Download the Common Core Curriculum Assessment Analysis template from the Digital Resource CD (filename: assessment.pdf).
2. Analyze the assessments in your curriculum according to the questions in the template.
3. Use your findings from the template to guide your search for additional assessment resources to help supplement the curriculum. Consider the following resources:
   - http://www.achievethecore.org
   - http://www.illustrativemathematics.org
   - http://www.insidemathematics.org
   - http://www.parcconline.org
   - http://www.smarterbalanced.org
   - You may also want to consider your state frameworks for support.

Comprehensive Assessment

Rigor is one of the three instructional shifts associated with today's standards. The three dimensions of conceptual knowledge, procedural fluency, and real-world applications are to be treated with equal intensity and balance. Assessments are intended to measure a student's meaningful understanding of mathematics. Meaningful understanding includes a comprehensive understanding across the three domains of rigor—procedural knowledge, conceptual knowledge, and real-world application-based knowledge. Incorporating the question types of constructed response, extended response, and performance tasks primarily target the development of conceptual knowledge and real-world application-based knowledge. As we begin to emphasize conceptual knowledge development, we must not lose sight of the fact that students must still develop procedural skills.

Procedural Fluency Assessment

Procedural fluency assessments can aid in the development of procedural knowledge. The word "fluently" appears in the content standards. It suggests accuracy and reasonable speed in computation. It is important that time is provided for students to develop grade-level expected fluencies alongside deep conceptual understanding.

The following grade level fluency assessments can be used as diagnostic tools, formative measures, and summative assessments of fluency. Examples of specific grade-level procedural fluency assessments can be located on the Digital Resource CD.

Figure 6.3 Sample Kindergarten Fluency Assessment

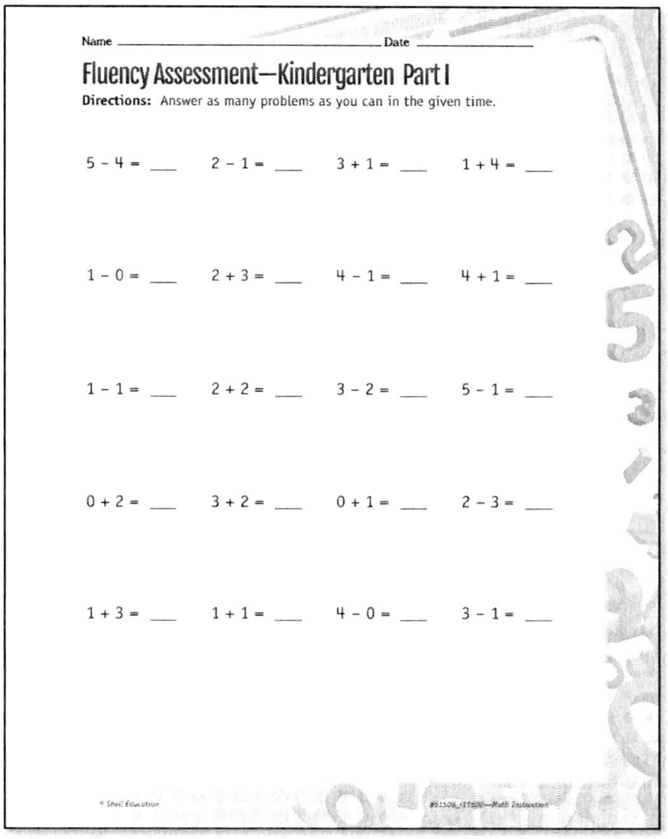
Name ______________________ Date ____________

Fluency Assessment—Kindergarten Part I

Directions: Answer as many problems as you can in the given time.

5 − 4 = ___	2 − 1 = ___	3 + 1 = ___	1 + 4 = ___
1 − 0 = ___	2 + 3 = ___	4 − 1 = ___	4 + 1 = ___
1 − 1 = ___	2 + 2 = ___	3 − 2 = ___	5 − 1 = ___
0 + 2 = ___	3 + 2 = ___	0 + 1 = ___	2 − 3 = ___
1 + 3 = ___	1 + 1 = ___	4 − 0 = ___	3 − 1 = ___

© Shell Education

Figure 6.4 Sample Grade 5 Fluency Assessment

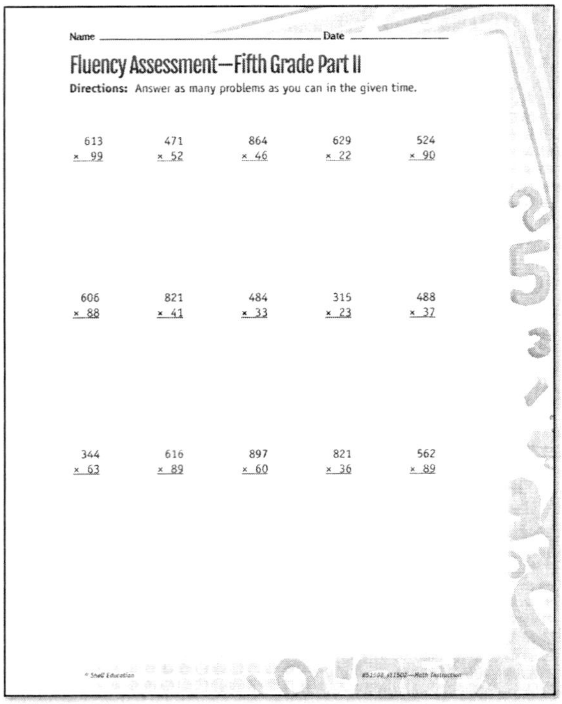

Name ______________________ Date __________

Fluency Assessment—Fifth Grade Part II

Directions: Answer as many problems as you can in the given time.

613×99	471×52	864×46	629×22	524×90
606×88	821×41	484×33	315×23	488×37
344×63	616×89	897×60	821×36	562×89

Figure 6.5 Sample Grade 8 Fluency Assessment

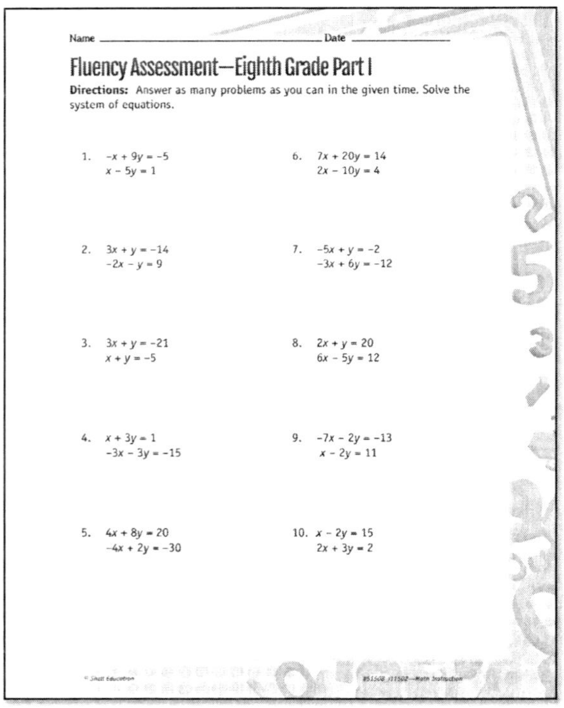

Name ______________________ Date __________

Fluency Assessment—Eighth Grade Part I

Directions: Answer as many problems as you can in the given time. Solve the system of equations.

1. $-x + 9y = -5$; $x - 5y = 1$
2. $3x + y = -14$; $-2x - y = 9$
3. $3x + y = -21$; $x + y = -5$
4. $x + 3y = 1$; $-3x - 3y = -15$
5. $4x + 8y = 20$; $-4x + 2y = -30$
6. $7x + 20y = 14$; $2x - 10y = 4$
7. $-5x + y = -2$; $-3x + 6y = -12$
8. $2x + y = 20$; $6x - 5y = 12$
9. $-7x - 2y = -13$; $x - 2y = 11$
10. $x - 2y = 15$; $2x + 3y = 2$

Stop and Think

Question 1

How might you ensure alignment between assessment expectations and what you are teaching in the classroom?

Question 2

How has your understanding of authentic assessments shifted as a result of reading this chapter? What beliefs have been reinforced?

Question 3

How might you incorporate the entire range of DOK level tasks into your assessment plan?

Conclusion

Effective mathematics instruction requires teachers to create meaningful learning environments that facilitate student learning. As the standards shift today, tomorrow, and in the future, what it means to know mathematics is constant. A student must have procedural knowledge, conceptual knowledge, factual knowledge, and metacognitive knowledge of mathematics. Students must have the ability to use this comprehensive knowledge to make sense of the world in which we live.

In order to help students develop their understanding of mathematics, the resources in this book are designed to help you develop your understanding of the mathematics standards. In turn, you will be able to design a curriculum that incorporates the shifts of focus, coherence, and rigor. You will be able to design authentic assessment tools that measure students' progress towards developing a comprehensive knowledge base.

Use the resources offered in this book to deliver effective mathematics instruction evidenced by students who will think and communicate as developing mathematicians.

Appendix A

References Cited

Barbeau, Edward J. 1989. "Mathematics for the Public." Paper presented at the International Commission on Mathematics Instruction. Leeds, England. September 1989.

Baroody, Arthur J. 1989. "One Point of View: Manipulatives Don't Come with Guarantees." *Arithmetic Teacher* 37 (2): 4–5.

Barr, Catherine, Doyle, Julie M. Clifford, Teresa De Leo, and Carol Dubeau. 2003. *"There is More to Math: A Framework for Learning and Math Instruction."* Waterloo Catholic District School Board CA Mathematics Framework. https://www.wcdsb.ca/programs/curriculum/math/pdf/math-developing-doc.pdf The Mathematics Framework was adopted by the California State Board of Education on November 6, 2013.

Berthold, Kristen, and Alexahdra Renkl. 2009. "Instructional Aids to Support a Conceptual Understanding of Multiple Representations." *Journal of Educational Psychology* 101 (1): 70–87.

California Department of Education. 2015. *Mathematics Framework Chapters*. http://www.cde.ca.gov/ci/ma/cf/draft2mathfwchapters.asp.

Canobi, Katherine H. 2009. "Concept-Procedure Interactions in Children's Addition and Subtraction." *Journal of Experimental Child Psychology* 102 (2): 131–149.

Canobi, Katherine H. 2004. "Individual Differences in Children's Addition and Subtraction Knowledge." *Cognitive Development* 18: 81–93.

Center for K–12 Assessment and Performance Management at ETS. Educational Testing Service position paper. "Coming Together to Raise Achievement: New Assessments for the Common Core State Standards." http://www.k12center.org/rsc/pdf/Coming_Together_April_2012_Final.PDF.

Clements, Douglas. H. 1999. "'Concrete' Manipulatives, Concrete Ideas." *Contemporary Issues in Early Childhood* 1 (1): 45–60.

Core Knowledge Foundation. 1999. *Core Knowledge Sequence: Content Guidelines for Grades K–8*. http://www.coreknowledge.org/mimik/mimik_uploads/documents/480/CKFSequence_Rev.pdf.

Daro, Phil. 2013. "Learning Trajectories in Mathematics." delivered at the California Mathematics Council Conference in Palm Springs, CA. November 2013

Freudenthal, Hans. 1991. *Revisiting Mathematics Education. China Lectures*. Dordrecht, Netherlands: Kluwer Academic Publishers.

Gredler, Margaret E. 2004. *Learning and Instruction: Theory Into Practice*. 5th ed. Upper Saddle River, NJ: Prentice Hall.

Hess, Karin K., Dennis Carlock, Ben Jones, and John R.Walkup. 2009. "What Exactly Do 'Fewer, Clearer, and Higher Standards' Really Look Like in the Classroom? Using a Cognitive Rigor Matrix to Analyze Curriculum, Plan Lessons, and Implement Assessments." In *Hess's Local Assessment Toolkit: Exploring Cognitive Rigor.* https://www.sde.idaho.gov/site/common/webinars/Cognitive%20Rigor%20Matrix%20Article_Hess,%20Carlock,%20Jones,%20and%20Walkup.pdf.

Hess, Karin K. 2006b. "Exploring Cognitive Demand in Instruction and Assessment." http://www.nciea.org/publications/DOK_ApplyingWebb_KH08.pdf.

Hiebert, James. 1986. *Conceptual and Procedural Knowledge: The Case of Mathematics*. Hillsdale, NJ: Erlbaum.

Hill, Heather C., Brian Rowan, and Deborah Lowenberg Ball. 2005. "Effects of Teachers' Mathematical Knowledge for Teaching on Student Achievement." *American Educational Research Journal* 42 (2): 371–406.

Joyce, Bruce, Marsha Weil, and Emily Calhoun. 2004. *Models of Teaching*. 7th ed. Boston, MA: Allyn & Bacon.

Kaplan, Sandra, and James Curry. 1985. *Curriculum Guide for the Education of Gifted High School Students*. Austin, TX: *Texas Association* for the *Gifted* and Talented (TAGT).

Lesh, Richard. 1990. "Computer-Based Assessment of Higher Order Understandings and Processes in Elementary Mathematics." In *Assessing Higher Order Thinking in Mathematics,* edited by Gerald Kulm, 81–110. Washington, DC: American Association for the Advancement of Science.

McCallum, Bill. 2011. "Structuring the Mathematical Practices." http://commoncoretools.me/2011/03/10/structuring-the-mathematical-practices/

McCallum, William. 2013. "Progressions Documents for the Common Core Math Standards." The University of Arizona Institute for Mathematics and Education. ime.math.arizona.edu/progressions.

Meyer, Anne, David H. Rose, and David Gordon. 2014. *Universal Design for Learning: Theory and Practice*. Wakefield, MA: CAST Professional Publishing.

National Governors Association (NGA) Center for Best Practices and Council of Chief State School Officers (CCSSO). 2010a. "Common Core State Standards: Mathematics." Washington, DC: www.corestandards.org.

National Governors Association (NGA) Center for Best Practices and Council of Chief State School Officers (CCSSO). 2010b. "Common Core State Standards for Mathematics." Washington, DC: www.corestandards.org.

Ohio Department of Education. 2015. "Common Core State Standards—Mathematics Standards Progressions." https://education.ohio.gov/getattachment/Topics/Ohio-s-New-Learning-Standards/Mathematics/Transition-Tools-Ohio-s-New-Learning-Standards-K-1/K-8-Standards-Progressions-2-14-12.pdf.aspx

Pressley, M., & Harris, K. R. (2006). Cognitive strategy instruction: From basic research to classroom instructions. In P. A. Alexander & P. H. Winne (Eds.), *Handbook of Educational Psychology* (2nd ed., pp. 265–287). Mahwah, NJ: Erlbaum.

Rittle-Johnson, Bethany, and Martha Wagner Alibali. 1999. "Conceptual and Procedural Knowledge of Mathematics: Does One Lead to the Other?" *Journal of Educational Psychology 91* (1): 175–189.

Star, Jon R. 2002. "Re-Conceptualizing Procedural Knowledge: The Emergence of 'Intelligent' Performances among Equation Solvers.'" In *Proceedings of the 28th Annual Meeting of the North American Chapter of the International Group for the Psychology of Mathematics Education*, edited by Denise Mewborn, Paolin Sztajn, Dorothy White, Heidi Wiegel, Robin Bryant, and Kevin Nooney, 999–1007. Columbus, OH: ERIC Clearinghouse for Science, Mathematics, and Environmental Education.

Star, Jon R., and Rittle-Johnson, Bethany. 2009. "It Pays to Compare: An Experimental Study on Computational Estimation." *Journal of Experimental Child Psychology* 102: 4 408–428.

Tatto, Maria Teresa, and Sharon Senk. 2011. "The Mathematics Education of Future Primary and Secondary Teachers: Methods and Findings from the Teacher Education and Development Study in Mathematics." *Journal of Teacher Education* 62 (2): 121–137.

Texas Association for the Gifted and Talented. 1991. "Universal Themes and Generalizations." Curriculum Guide for the Education of Gifted High School Students. http://www.ggusd.us/ourpages/gate/teacher_res//universal%20themes/UNIVERSALTHEMESANDGENERALIZATIONS.pdf.

Tomlinson, Carol Ann. 2000. "Differentiation of Instruction in the Elementary Grades." ERIC Digest. ERIC Clearinghouse on Elementary and Early Childhood Education.

TN Core. 2015. "Common Core State Standards, Fluency in Mathematics." Accessed June 12, 2015 http://tncore.org/sites/www/Uploads/2.25.13Additions/fluency%20documents%20final.pdf.

Tyler, Ralph W. 1949. *Basic Principles of Curriculum and Instruction*. Chicago, IL: The University of Chicago Press.

US Department of Education. 2001. *No Child Left Behind Act of 2001*. Washington DC. http://www2.ed.gov/policy/elsec/leg/esea02/index.html.

Van Zoest, Laura. R., Shari L. Stockero, and Cynthia E. Taylor. 2011. "The Durability of Sociomathematical Norms Intentionally Fostered in an Early Pedagogy Course." *Journal of Mathematics Teacher Educators* doi:10.1007/s10857-011-9183-y.

Webb, Norman. 1999. "Alignment of Science and Mathematics Standards and Assessments in Four States. Research Monograph No. 18." Washington, DC: Council of Chief State School Officers.

Appendix B

Additional Resources

Common Core Vocabulary List

The following vocabulary list represents words from the Common Core State Standards for mathematics. The emphasis on academic language and attention to precision calls upon the specific and technical language of mathematics. Students are charged with constructing viable arguments. These arguments are highly dependent upon their command of the language of mathematics.

The grade-level lists presented also show the location of the word in the Common Core standards. These lists can be used to create word walls, student vocabulary journals (define and illustrate), and assessments (students required to use in context).

The vocabulary is written in a grade K–8 matrix to emphasize the coherent progression. Teachers are encouraged to understand the interdependence of content across grade levels.

The entire list of vocabulary words can be located on the Digital Resource CD.

Common Core Mathematics Vocabulary List—Coherence

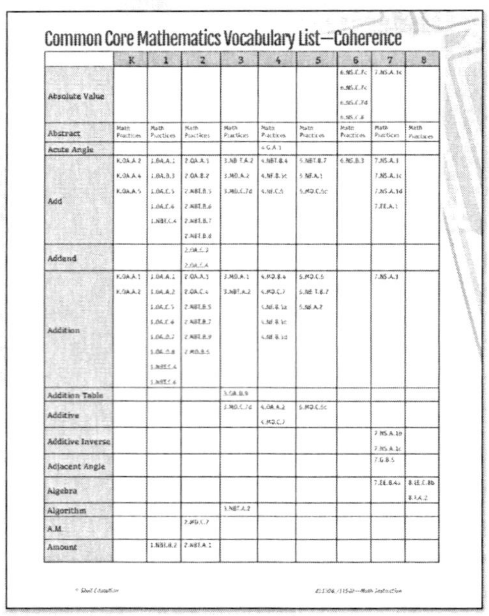

Common Core Mathematics Vocabulary List—Coherence

	K	1	2	3	4	5	6	7	8
Absolute Value							6.NS.C.7c 6.NS.C.7c 6.NS.C.7d 6.NS.C.8	7.NS.A.1c	
Abstract	Math Practices	Math Practices	Math Practices	Math Practices	Math Practices	Math Practices	Math Practices	Math Practices	Math Practices
Acute Angle					4.G.A.1				
Add	K.OA.A.2 K.OA.A.4 K.OA.A.5	1.OA.A.1 1.OA.B.3 1.OA.C.5 1.OA.C.6 1.NBT.C.4	2.OA.A.1 2.OA.B.2 2.NBT.B.5 2.NBT.B.6 2.NBT.B.7 2.NBT.B.8	3.NBT.A.2 3.MD.A.2 3.MD.C.7d	4.NBT.B.4 4.NF.B.3c 4.NF.C.5	5.NBT.B.7 5.NF.A.1 5.MD.C.5c	6.NS.B.3	7.NS.A.1 7.NS.A.1c 7.NS.A.1d 7.EE.A.1	
Addend			2.OA.C.3 2.OA.C.4						
Addition	K.OA.A.1 K.OA.A.2	1.OA.A.1 1.OA.A.2 1.OA.C.5 1.OA.C.6 1.OA.D.7 1.OA.D.8 1.NBT.C.4 1.NBT.C.6	2.OA.A.1 2.OA.C.4 2.NBT.B.5 2.NBT.B.7 2.NBT.B.9 2.MD.B.5	3.MD.A.1 3.NBT.A.2	4.MD.B.4 4.MD.C.7 4.NF.B.3a 4.NF.B.3c 4.NF.B.3d	5.MD.C.5 5.NBT.B.7 5.NF.A.2		7.NS.A.1	
Addition Table				3.OA.B.9					
Additive				3.MD.C.7d	4.OA.A.2 4.MD.C.7	5.MD.C.5c			
Additive Inverse								7.NS.A.1b 7.NS.A.1c	
Adjacent Angle								7.G.B.5	
Algebra								7.EE.B.4a	8.EE.C.8b 8.F.A.2
Algorithm				3.NBT.A.2					
A.M.			2.MD.C.7						
Amount		1.NBT.B.2	2.NBT.A.1						

© Shell Education

Online Resources

The following represent a few online resources to support Common Core implementation.

Achieve the Core

www.achievethecore.org

This site was developed by lead writers of the Common Core standards. It provides tasks, lessons, and professional development modules.

The Common Core Standards Initiative

www.corestandards.org

This is the home site of the Common Core State Standards. State specific additions are also included.

EngageNY

www.engageny.org

This site provides lessons, unit modules, and professional development resources for teachers.

Illustrative Mathematics

www.illustrativemathematics.org

This site provides a range of question types that can be downloaded. This site also provides direct guidance to the testing consortia and state curriculum developers.

Inside Mathematics

www.insidemathematics.org

Professional mathematics resource that includes classroom videos, tasks, and assessment tools.

Math Assessment Project (MAP)

www.map.mathshell.org

This site provides grades 6–12 performance tasks, lessons, and professional development resources for educators.

Partnership for Assessment of Readiness for College and Careers

www.parcconline.org

This is the home site for the PARCC assessment system. The site includes the most recent updates relating to PARCC and educational resources.

Smarter Balanced Assessment Consortium

www.smarterbalanced.org

This is the home site for the SBAC assessment system. The site includes the most recent updates and news relating to SBAC and educational resources. A practice exam is also available.

Tools for the Common Core State Standards in Math

www.commoncoretools.me

This is a blog by William McCallum, one of the primary authors of the Common Core, and provides tools and resources for implementation.

Appendix C

Contents of the Digital Resource CD

Chapter 1 Resources

Page(s)	Title	Filename
21	Common Core Domain Coherence	coherence.pdf
N/A	Content Standard Analysis Chart—Kindergarten	contentkin.pdf contentkin.docx
N/A	Content Standard Analysis Chart—First Grade	contentone.pdf contentone.docx
N/A	Content Standard Analysis Chart—Second Grade	contenttwo.pdf contenttwo.docx
N/A	Content Standard Analysis Chart—Third Grade	contentthree.pdf contentthree.docx
N/A	Content Standard Analysis Chart—Fourth Grade	contentfour.pdf contentfour.docx
23	Content Standard Analysis Chart—Fifth Grade	contentfive.pdf contentfive.docx
N/A	Content Standard Analysis Chart—Sixth Grade	contentsix.pdf contentsix.docx
N/A	Content Standard Analysis Chart—Seventh Grade	contentseven.pdf contentseven.docx
N/A	Content Standard Analysis Chart—Eighth Grade	contentseven.pdf contentseven.docx
25	Cluster Analysis Chart Template	cluster.pdf cluster.docx
31	Mathematical Practice Analysis Tool	analysis.pdf analysis.docx

35	Content and Practices Alignment Tool	alignment.pdf alignment.docx
36	Visual Prompts	visualprompts.pdf
46	Question Frame Template	qframe.pdf

Chapter 2 Resources

Page(s)	Title	Filename
60–61	Curriculum Focus Audit Tool—Kindergarten	audittoolk.pdf audittoolk.docx
N/A	Curriculum Focus Audit Tool—First Grade	audittoolone.pdf audittoolone.docx
N/A	Curriculum Focus Audit Tool—Second Grade	audittooltwo.pdf audittooltwo.docx
N/A	Curriculum Focus Audit Tool—Third Grade	audittoolthree.pdf audittoolthree.docx
N/A	Curriculum Focus Audit Tool—Fourth Grade	audittoolfour.pdf audittoolfour.docx
N/A	Curriculum Focus Audit Tool—Fifth Grade	audittoolfive.pdf audittoolfive.docx
N/A	Curriculum Focus Audit Tool—Sixth Grade	audittoolsix.pdf audittoolsix.docx
N/A	Curriculum Focus Audit Tool—Seventh Grade	audittoolseven.pdf audittoolseven.docx
N/A	Curriculum Focus Audit Tool—Eighth Grade	audittooleight.pdf audittooleight.docx

Chapter 3 Resources

Page(s)	Title	Filename
69	Common Core State Standards Progressions Charts	progressions.pdf
71–72	Cluster Analysis Tool Template	clustertool.pdf clustertool.docx
72	K–8 Cluster Emphasis Chart	clusteremph.pdf

Page(s)	Title	Filename
N/A	Focus and Coherence Analyzer Template	analyzetemp.pdf analyzetemp.docx
73	Focus and Coherence Analyzer—Kindergarten	analyzerkinder.pdf analyzerkinder.docx
75	Focus and Coherence Analyzer—First Grade	analyzerone.pdf analyzerone.docx
77	Focus and Coherence Analyzer—Second Grade	analyzertwo.pdf analyzertwo.docx
79	Focus and Coherence Analyzer—Third Grade	analyzerthree.pdf analyzerthree.docx
81	Focus and Coherence Analyzer—Fourth Grade	analyzerfour.pdf analyzerfour.docx
83	Focus and Coherence Analyzer—Fifth Grade	analyzerfive.pdf analyzerfive.docx
85	Focus and Coherence Analyzer—Sixth Grade	analyzersix.pdf analyzersix.docx
87	Focus and Coherence Analyzer—Seventh Grade	analyzerseven.pdf analyzerseven.docx
89	Focus and Coherence Analyzer—Eighth Grade	analyzereight.pdf analyzereight.docx
N/A	Focus and Coherence Planner—Kindergarten	plannerkinder.pdf plannerkinder.docx
N/A	Focus and Coherence Planner—First Grade	plannerone.pdf plannerone.docx
N/A	Focus and Coherence Planner—Second Grade	plannertwo.pdf plannertwo.docx
91–93	Focus and Coherence Planner—Third Grade	plannerthree.pdf plannerthree.docx
N/A	Focus and Coherence Planner—Fourth Grade	plannerfour.pdf plannerfour.docx
N/A	Focus and Coherence Planner—Fifth Grade	plannerfive.pdf plannerfive.docx
N/A	Focus and Coherence Planner—Sixth Grade	plannersix.pdf plannersix.docx

Page(s)	Title	Filename
N/A	Focus and Coherence Planner—Seventh Grade	plannerseven.pdf plannerseven.doc
N/A	Focus and Coherence Planner—Eighth Grade	plannereight.pdf plannereight.docx
108	Universal Concept Matrix	matrix.pdf matrix.docx
109	Coherence Ladder Template	coladder.pdf
110	Coherence Cube	cocube.pdf
113	Math Chain Template	mathchain.pdf mathchain.docx
114	Coherence Puzzle Template	copuzzle.pdf

Chapter 4 Resources

Page(s)	Title	Filename
126	Rigor Calibration Tool	calibration.pdf
127	Rigor Analysis Tool	riganalysis.pdf riganalysis.docx
N/A	Rigor Monitoring Tool—Kindergarten	monitoringk.pdf monitoringk.docx
N/A	Rigor Monitoring Tool—First Grade	monitoringone.pdf monitoringone.docx
129–131	Rigor Monitoring Tool—Second Grade	monitoringtwo.pdf monitoringtwo.docx
N/A	Rigor Monitoring Tool—Third Grade	monitoringthree.pdf monitoringthree.docx
N/A	Rigor Monitoring Tool—Fourth Grade	monitoringfour.pdf monitoringfour.docx
N/A	Rigor Monitoring Tool—Fifth Grade	monitoringfive.pdf monitoringfive.docx
N/A	Rigor Monitoring Tool—Sixth Grade	monitoringsix.pdf monitoringsix.docx

N/A	Rigor Monitoring Tool—Seventh Grade	monitoringseven.pdf monitoringseven.docx
N/A	Rigor Monitoring Tool—Eighth Grade	monitoringeight.pdf monitoringeight.docx

Chapter 5 Resources

Page(s)	Title	Filename
139–140	FCR Unit Frame Protocol	fcr.pdf
143–144	Inquiry-Based Math Lesson Template	inquirybase.pdf inquirybase.docx
144	Meaningful Mathematics (M^2) Criteria Chart	meaningfulmath.pdf
1465	Perpetual Inquiry Chart	perpetual.pdf
N/A	Direct Instruction Math Lesson Template	directinstruct.pdf directinstruct.docx
N/A	Mathematics Strategies Template	strategies.pdf strategies.docx

Chapter 6 Resources

Page(s)	Title	Filename
208	Common Core Curriculum Assessment Analysis Template	assessment.pdf assessment.docx
N/A	Fluency Assessment Tracking Sheet	fluencytrack.pdf
209	Fluency Assessment—Kindergarten I and II	fluencykinderi.pdf fluencykinderii.pdf
N/A	Fluency Assessment—First Grade I and II	fluencyonei.pdf fluencyoneii.pdf
N/A	Fluency Assessment—Second Grade I and II	fluencytwoi.pdf fluencytwoii.pdf
N/A	Fluency Assessment—Third Grade I and II	fluencythreei.pdf fluencythreeii.pdf
N/A	Fluency Assessment—Fourth Grade I and II	fluencyfouri.pdf fluencyfourii.pdf

Page(s)	Title	Filename
210	Fluency Assessment—Fifth Grade I and II	fluencyfivei.pdf fluencyfiveii.pdf
N/A	Fluency Assessment—Sixth Grade I and II	fluencysixi.pdf fluencysixii.pdf
N/A	Fluency Assessment—Seventh Grade I and II	fluencyseveni.pdf fluencysevenii.pdf
210	Fluency Assessment—Eighth Grade I and II	fluencyeighti.pdf fluencyeightii.pdf

Appendix B Resources

Page(s)	Title	Filename
216	Common Core Mathematics Vocabulary List—Coherence	vocabulary.pdf

Notes